Modern Apartment Design

Modern Apartment Design provides guidelines to the design of modern apartment buildings as well as a summation of current cutting-edge practice in engineered timber construction.

The book covers a brief history of apartment buildings around the world, with a broad outline of different types of apartment blocks. It has a strong focus on the design and actual construction of apartment buildings, especially those utilising mass timber, such as cross-laminated timber and laminated veneer lumber. It also features six Case Study chapters from industry-leading practitioners in the area, enabling best practice in architecture and engineering of these new apartment building types to be more widely understood and propagated worldwide.

The fully illustrated, full-colour case studies span the globe and include: Clearwater Quay in Christchurch, New Zealand (Pacific Environments NZ); Wynyard Central East 2 in Auckland, New Zealand (Architectus); Dalton Works in London, UK (Waugh Thistleton Architects); Mjøstårnet, Brumunddal, Norway (Voll Arkitekter); Brock Commons Tallwood House student housing in Vancouver, Canada (Acton Ostry Architects); and Regensbergstrasse apartments in Zurich, Switzerland (Dreicon). The book will be of great interest to architects and architecture students.

Guy Marriage is an architect, registered in Britain and in New Zealand. Now teaching construction, design and building science at Victoria University of Wellington in New Zealand, he gained construction experience from over two decades in practice, with 11 years working in architectural practices in London including Colman Architects, Jestico + Whiles, and Foster + Partners. Returning to New Zealand in 2000, he worked for 7 years at the Studio of Pacific Architecture and was one of the founding directors of First Light Studio in 2012. He has experience in projects that range in scale from small homes, prefabricated houses, and apartment buildings, as well as shopping centres, multi-storey office buildings, and two of the largest Jubilee Line underground railway stations in London. In 2019, he published his first book *Tall: The Design and Construction of High-Rise Architecture*.

For further details, please visit: www.tallconstruction.org

Modern Apartment Design

Author: Guy Marriage

Contributing authors: Dr Nabil Jose Allaf, Pamela Bell, Emily Newmarch, Dr Emina Kristina Petrović, Dr Sara Shabahang, Joseph Sturm, Russell Acton, Guy Marriage, Patrick Clifford, Joe Kaps, Roger French, Øystein Elgsaas, Kieran Walker, Michael Newcombe.

Contributing architects: Acton Ostry Architects, Architectus, Dreicon, Pacific Environments NZ, Voll Arkitekter, Waugh Thistleton Architects, Enovate.

Illustrations: Guy Marriage and Lauren Hayes, apart from Case Study chapters, listed below:

Clearwater Quays: Pacific Environments NZ

Wynyard Central: Architectus and photos by Simon Devitt

Dalston Works: Waugh Thistleton Architects and photos by Daniel Shearing

Mjøstårnet: Voll Arkitekter

Brock Commons: Acton Ostry Architects

Regensbergstrasse: DREICON

Edited by Guy Marriage

Routledge
Taylor & Francis Group
LONDON AND NEW YORK

First published 2022
by Routledge
2 Park Square, Milton Park, Abingdon, Oxon OX14 4RN

and by Routledge
605 Third Avenue, New York, NY 10158

Routledge is an imprint of the Taylor & Francis Group, an informa business

British Library Cataloguing-in-Publication Data
A catalogue record for this book is available from the British Library

Library of Congress Cataloging-in-Publication Data
Names: Marriage, Guy, editor.
Title: Modern apartment design / edited by Guy Marriage.
Description: Abingdon, Oxon; New York: Routledge, 2022. | Includes bibliographical references and index.
Identifiers: LCCN 2021018967 (print) | LCCN 2021018968 (ebook) | ISBN 9780367643096 (hardback) | ISBN 9780367643102 (paperback) | ISBN 9781003123873 (ebook)
Subjects: LCSH: Apartment houses—Design and construction. | Engineered wood construction. | Apartment houses—Design and construction—Case studies. | Engineered wood construction—Case studies.
Classification: LCC TH4820 .M59 2022 (print) | LCC TH4820 (ebook) | DDC 728/.314—dc23
LC record available at https://lccn.loc.gov/2021018967
LC ebook record available at https://lccn.loc.gov/2021018968

ISBN: 978-0-367-64309-6 (hbk)
ISBN: 978-0-367-64310-2 (pbk
ISBN: 978-1-003-12387-3 (ebk)

DOI: 10.4324/9781003123873

Typeset in Univers LT Std
by codeMantra

Cover photograph: Wynyard Central: building by Architectus, photograph by Simon Devitt.

Contents

Preface

I started writing this book in early March 2020, flying in to Canada on a cold, sharp, bright spring morning; far-off mountains capped with ice and snow geese honking overhead. The streets were full of smiling, laughing people, the malls were packed with shoppers, and building sites were busy with the construction of new high-rise apartment buildings for eager Vancouverites. In far-off lands, a small corona virus was causing havoc. The official travel advice when I set off on my research trip was that you should be cautious, but there were no travel restrictions apart from three countries: China, Iran, and Italy.

Vancouver is beautiful in spring: the blossoms are opening on the cherry trees, while the snow-clad mountains remind you that winter was not long gone. All over suburban Vancouver, clumps of apartment buildings are springing up, mainly clustered around transport hubs provided by the Sky Train. As Vancouver and Richmond lie on the flat Fraser River delta, the effect is very easy to see: rapid public transport station + viable apartment site = flourishing urban life. Suburban lifestyle, by contrast, demands a car for transport or at the very least, a nearby bus stop. The friends I stayed with in Vancouver live in pleasant suburban surroundings with a view out to sea, right on the edge of the Strait of Georgia. Sixteen thousand kilometres away, on the opposite side of the Pacific Ocean, lays my home in Aotearoa New Zealand.

Very soon, as we all know now, the world turned upside down and rapidly ground to a halt.

A mere seven days after my arrival in Canada, shops were shutting in the mall, people were being openly shunned in the street, and the world turned into a deathly version of the Zombie Apocalypse before my eyes. As the Covid-19 pandemic unfolded, first one country then another turned up with evermore frightening statistics, and increasingly, deaths of the elderly and the unwell. In this strange new world, who knew who would live, and who would die? That may seem melodramatic now, but early reported statistics (like 3.4% of cases dying) meant that while young people were surviving, the elderly and unwell were bearing the brunt of the casualties: nearer 9% deaths were from the elderly. It's hard to stay in a pleasant mood with someone who sneezes on you when there is a one-in-twenty chance that this single sneeze may end up killing you or, more likely, you pass this death sentence onto your parents and grand-parents.

Exactly what effect this virus will have on the future of the built environment is completely unknown at this stage, but self-isolation and social distancing (new concepts in March 2020) can work just as well in suburbia as they can in an urban situation. My interest is primarily in whether the pandemic will have an effect on where we live and work, now and in the future. This book sets out to explore the world of dense, urban apartment living and its future desirability in a post-Covid world.

As I flew out of Vancouver to hurry back to the safe bolt-hole of Aotearoa, largely Covid-free, the snow goose continued to fly in their V-formations, oblivious to the issues of the humans hiding in lockdown below.

Guy Marriage
Pukerua
Te Upoko o te Ika a Maui
Aotearoa, 2021.

Acknowledgements

As with any book project, there are a number of people to acknowledge for their help. First and foremost are Sue Gall, Paul Wright, Quinn Wright, and Alice Wright for their help and assistance in Vancouver. Also to be thanked is Jacque Gilbert in Amsterdam: next time, OK?

Mary Charleson from Charleson Communications, for her timely advice and help. Thanks to Greg Johnson at UBC SALA, for useful advice on the state of Canadian apartment construction, and thanks too to Janice Wong for her timely and valuable advice on New York apartment design. Huge thanks to Meret Gabra, David Liddell, and Maya Gabra-Liddell for their continued support and advice on residential conditions in the UK.

I particularly want to thank Alice Masters for her heroic last-minute proof-reading, and to thank Vanessa Coxhead for providing impetus for this book and the last one. Heike Gerhards deserves a reward for her sourcing of German noise-cancelling administration, while Morten Gjerde also receives thanks for advice on Norwegian building conditions, and helping get the Norwegian project on board. Thanks too to Brittany Irvine for a great cover design, and thanks to Simon Devitt for a great cover photograph!

Most of all, I want to thank my mother Robin Marriage, for unfailingly having faith in me that this project could be completed and that the end result would be worth reading. Thanks Mum.

Huge thanks go once again to the illustrator of much of this book, Lauren Hayes, who has again done a great job. She turned my scribbles and sketches into good-looking, meaningful drawings. My thanks also go to the co-sponsors of the Summer Scholarship program at Victoria University of Wellington (Te Herenga Waka) that helped create this book: Victoria University and First Light Studio. Thank you all, especially to my fellow directors at First Light, and my colleagues at Victoria University, who all contribute in their own special ways.

The class of the Third-Year Construction paper at Victoria University over a number of years deserve special thanks, for inspiring both this book and the last book.

Finally, my sincere thanks to all the numerous contributors to this book from all around the world, for putting together their writings and illustrations during the biggest global pandemic of the last century. Architects, clients, engineers, and contractors alike: we're all focused on bringing the best apartment buildings to life. Thank you so much – I really could not have done it without you all. He aroha whakatō, he aroha puta mai.

orchid.org/0000-0002-9794-5451

Abbreviations

ACM	Aluminium Composite Material
ACP	Aluminium Composite Panel
BBS	Binderholz Bausysteme
BIM	Building Information Model
BMU	Building Maintenance Unit
BRDF	Bioenergy Research Demonstration Facility
BREEAM	Building Research Establishment Environmental Assessment Method
CCSCR	Climate Change and Sea Level Rise
CH_4	Methane
CLT	Cross-Laminated Timber
CO_2	Carbon Dioxide
Covid	(or COVID-19) Coronavirus disease that started global circulation in 2019
CR2	Cross Rail 2
CTBUH	Council on Tall Buildings and Urban Habitat
DB	Distribution Board
DES	District Energy System
DLT	Dowell-Laminated Timber
DPC	Damp Proof Course
EIFS	Externally Insulated Façade System
EPD	Environmental Product Declarations
ETICS	External Thermal Insulation Composite System
FRR	Fire Resistance Rating
GLT	Glue-Laminated Timber
GOD	Greenspace Oriented Development
GWP	Global Warming Potential
GHG	Greenhouse gases
HPL	High-Pressure Laminate
HS1	High Speed 1
HVAC	Heating, Ventilation, Air-conditioning
IGU	Insulated Glazing Unit
IPCC	International Panel on Climate Change
ISO	International Standards Organisation
LCA	Life Cycle Analysis
LED	Light-Emitting Diode
LEED	Leadership in Energy and Environmental Design
LSF	Light Steel Frame
LTF	Light Timber Frame

LVL	Laminated Veneer Lumber
MEP	Mechanical, Electrical, Plumbing
MMC	Modern Methods of Construction
MTC	Mass Timber Construction
NIWA	National Institute of Water and Atmospheric Research
NLT	Nail-Laminated Timber
NZSO	New Zealand Symphony Orchestra
OSB	Oriented Strand Board
OSM	Off-Site Manufacturing
PIR	Polyisocyanurate
PMVC	Pre-manufactured Volumetric Construction
PPM	parts per million
PRESSS	Precast Seismic Structural System
PSL	Parallel Strand Lumber
PV	Photo Voltaic
PVC	Poly Vinyl Chloride
QA	Quality Assurance
RAB	Rigid Air Barrier
RFID	Radio Frequency Identification
SE	Structural Engineer
SARS	Severe Acute Respiratory Syndrome
SDE	Shared Data Environment
SLS	Serviceability Limit State
SPD	Staircase Pressurisation Duct
SWB	Switch Board
TOD	Transport Oriented Development
ULS	Ultimate Limit State
VOC	Volatile Organic Compounds
WQ	Wynyard Quarter

Glossary

Glossary All definitions from Harris, Cyril. 2006. *Dictionary of Architecture and* Construction (*4th edition*). New York: McGraw Hill.

apartment A room or suite of rooms designed to be lived in, containing at least one bathroom; is separated from, and is usually one of, many similar units within a multiple dwelling.

atrium The main hall of an ancient Roman house, containing an opening to the sky (compluvium) through which rainwater falls to a tank or cistern below (impluvium). **2.** In a contemporary building, a large vertical space, often centrally located, that connects three or more floors and creates a sense of spaciousness.

courtyard An open area that is partially or fully enclosed by one or more buildings and/or by walls. Courtyards that are enclosed or partially enclosed by walls are sometimes referred to as patios.

condominium A form of real estate ownership of a multifamily residential dwelling. Each occupant has 100% ownership of [their] own apartment and partial ownership of common elements such as hallways, elevators, plumbing, etc.

co-op An apartment building collectively owned by a corporation, where residents buy shares in the corporation and lease their apartment under a cooperative agreement.

corridor **1.** A long interior passageway providing access to several rooms. **2.** A public means of access from several rooms or spaces to an exit. **3.** An enclosed passageway that limits the means of egress to a single path of travel.

detached house A house that stands completely alone, not sharing a wall with another house.

duplex apartment A separate dwelling in an apartment building, having rooms on two levels, with self-contained vertical circulation.

enfilade The alignment of a series of doors axially through a sequence of rooms.

flat One floor of a multi-storey building or a dwelling unit on one floor.

insula **1.** In Roman town planning, a block of buildings surrounded by streets. **2.** A Roman apartment house occupying such a block.

lanai A living room or lounge area which is entirely, or in part, open to the outdoors.

maisonette see **duplex apartment**.

mansard roof (*US* and *Brit.*) A roof having a double slope on all four sides, the lower slope being much steeper.

mansion **1.** A very large, imposing, stately residence. **2.** In colonial times, the residence of a landholder. **3.** A **manor house**; also called a mansion house.

partition A dividing wall within a building; may be bearing or non-load-bearing.

party wall A wall used jointly by two parties under easement agreement, erected upon a line dividing two parcels of land, each of which is a separate real estate entity; a common wall.

(also known as a demising wall).

row house One of an unbroken line of houses sharing one or more side-walls with its neighbours.

semi-detached (Brit) aka **double house** (US) **1.** A pair of **semi-detached houses** having a plan that is symmetric on both sides of the common wall; each of the two units has its own entry.

suburb An outlying area, in or near a city, of predominantly residential land use.

terrace house One of a row of houses situated on a **terrace**, or similar site.

townhouse **1.** A comfortable-to-luxurious dwelling in an urban environment. **2.** One of a series of houses constructed in an unbroken row, separated by **party walls**, often with a relatively flat roof. **3.** An upscale **row house**.

triplex house A house that provides living quarters for three families, each with a separate entrance; usually has three stories, with one apartment on each floor.

urban area An area which is within the city limits, or closely linked to it by common use of public utilities or services.

urban sprawl An unplanned development of open land, usually on the outskirts of a city.

vestibule An **anteroom** or small foyer leading into a larger space.

Contributors

Guy Marriage

Guy Marriage is an architect, registered in Britain and in New Zealand. Now teaching construction, design and building science at Victoria University of Wellington in New Zealand, he gained construction experience from over two decades in practice, with 11 years working in architectural practices in London including Colman Architects, Jestico + Whiles, and Foster + Partners. Returning to New Zealand in 2000, he worked for 7 years at the Studio of Pacific Architecture and was one of the founding directors of First Light Studio in 2012. He has experience in projects that range in scale from small homes, prefabricated houses, and apartment buildings, as well as shopping centres, multi-storey office buildings, and two of the largest Jubilee Line underground railway stations in London. In 2019, he published his first book *Tall: The Design and Construction of High-Rise Architecture*.

Lauren Hayes

Lauren Hayes is an architectural graduate at First Light Studio and tutor of architecture at Victoria University of Wellington, with an adept hand at drafting many of the illustrations you see in this book. Her passions include sustainable design and contemporary residential architecture, which she has pursued both in New Zealand and abroad at Eindhoven University of Technology in the Netherlands. Her recently completed Master's research focuses on the global sand shortage and a discussion between poetic and scientific sustainable design methods.

Dr Nabil Jose Allaf

Dr Nabil Jose Allaf studied architecture at the National School of Architecture in Nantes. After working as a practicing architect in France, he completed his PhD under supervision of Associate Professor Andrew Charleson at Victoria University in New Zealand. His PhD work is a seminal work on the integration of seismic retrofit and architecture in existing unreinforced masonry buildings, approaching seismic structures through their capacity to enrich architectural qualities. He is currently teaching in both France and New Zealand.

Adjunct Professor Pamela Bell

Pamela Bell brings experience in Architecture research (MArch, *Kiwi Prefab* book + exhibition), Olympic-level sport (snowboarding at Winter Olympics, Japan), business entrepreneurship (Prefab NZ, Inc., Fruition Apparel Ltd, NZ Snowboard Academy Ltd), and governance (Urban Plus, Brosnan Group, ConCOVE, MBIE Building Advisory Panel, BRANZ, Mt Vic Hub). She founded the peak body for innovative construction, PrefabNZ, in 2010 from her Master of Architecture thesis *Kiwi Prefab*, and handed over the reins in 2019 when offsite construction in New Zealand reached a tipping point of awareness for central government, the construction industry and the public consumer alike.

Today, Pamela is an independent Innovation Consultant, specialising in the built environment, growing start-ups and future-focused governance to clients around the world. She has been the lucky recipient of a number of awards throughout her career, most recently the Victoria University of Wellington Alumni Award (2019) and an honorary appointment as Adjunct Professor (2020).

Emily Newmarch

Emily is a PhD Candidate at Victoria University of Wellington in the School of Architecture exploring how measuring embodied carbon can be integrated into the design of low-carbon commercial buildings in New Zealand. Prior to undertaking this research, Emily has worked in small and large architecture practices, and completed a Master's thesis on the thermal performance in residential buildings with an exploration into the poetic relationship between architecture and environment. Emily has also taught in the fields of environmental science, construction, architectural design, and research methodologies.

Dr Emina Kristina Petrović

Dr Emina Kristina Petrović is a Senior Lecturer at the Wellington School of Architecture, Victoria University of Wellington, where she is also the Postgraduate Programmes Director. Her book *Materials for Healthy, Ecological and Sustainable Built Environment: Principles for Evaluation* (2017, co-authored with Brenda Vale and Maibritt Pedersen Zari) demonstrates Petrović's research focus on the impacts building materials can have on human health in the indoor environment. Petrović is deeply committed to the great need of sustainable change and that an active rethinking of the built environment plays a critical role in this process, including housing forms and preferences.

Dr Sara Shabahang

Dr Sara Shabahang studied architecture at the Azad University of Mashhad. After working as a head architect and lecturer in Iran, she completed her PhD under the supervision of Professor Brenda Vale at Victoria University in New Zealand. Her PhD is about the statistical assessment of neighbourhood characteristics on air and surface temperatures, and was done through a simulation method using ENVI-met software. She is currently teaching at Victoria University of Wellington and working as an architect in New Zealand.

Joseph Sturm

Joseph Sturm is a building scientist based in Tallinn, Estonia with expertise in medium density housing, project management, and sustainability. He specialises in international projects and has worked as a project analyst for a Swiss-Estonian prefab start-up as well as a project manager in a leading architectural practice. Originally from New Zealand and trained at Victoria University of Wellington, Joseph's professional interests also include design competitions, lessons-learned processes, building and planning regulations, housing assessment metrics, and definitions of the term liveability. He is an enthusiastic proponent of urban intensification.

Acton Ostry Architects

Acton Ostry Architects (AOA) is a leading Canadian design practice recognised for the creation of eloquent and innovative community, public institutional, mixed-use residential, and tall wood buildings. Their experience encompasses a wide range of facility programming, feasibility studies, master planning, rezoning, architectural design, and interior design services. They are members of the Canada Green Building Council and the Canadian Passive House Institute. Led by Russell Acton and Mark Ostry, the practice of 35 architectural and administrative staff has a well-established track record for delivering award-winning architecture, demonstrated by their receipt of over 100 local, national, and international awards.

Architectus

Patrick Clifford and Severin Soder are both Principals of Aotearoa architecture and urbanism practice, Architectus. As one of the most successful and critically acclaimed practices in New Zealand, Architectus has been realising innovative and bespoke projects throughout the country for the past 35 years. With an emphasis on people and sustainability, the practice's work has been recognised for its clarity of intent and resolution, its careful reading of context and content and ability to advance a broad cultural agenda.

Voll Arkitekter

Øystein Elgsaas is the principal architect at Voll Arkitekter in Trondheim, Norway. With 23 employees, ranging from experienced architects and planners with broad expertise and young employees with a fresh approach, Voll undertakes all types of architecture design.

Voll always strive to take the project one step closer to realisation. Their vision is that every project they do, should be built. They emphasise quality, realism, and feasibility in all project phases. Voll undertake projects all over Norway, and their clients are from both the public and private sectors. Locally Voll have had several high-profiled projects, but Mjøstårnet is the first to gain global attention.

Pacific Environments

Pacific Environments Architects is an environmentally considerate architectural practice focusing on the best outcomes for people, place, and community. They believe that designing physical objects, the built environment and associated infrastructure to comply with the principles of social, economic, and ecological sustainability are core to a constantly regenerating and self-sustaining ecosystem. As signatories to Architects Declare New Zealand, they are committed to strengthening the working process to create architecture and urbanism that has a more positive impact on the world around us.

DREICON

DREICON was founded in Switzerland by Niels Hochuli in 2010, with Joe Kaps joining the practice in 2011. The two partners combine over 50 years of experience in the building industry. Their very different but complementary strengths make them a well-balanced team with unique project know-how for very complex projects. DREICON is now a team of 12, working both in Switzerland and internationally.

For ten years, DREICON have pursued a uniquely bespoke architecture based on a well-crafted balance between their client's individual needs and the opportunities forthcoming from their projects' context. They apply this pursuit of their core principals to all of their projects. The diversity in architectural language and expression is a product of a very intensive collaborative analysis process with the clients.

DREICON's invaluable project management experience accompanies each of their projects from outset to conclusion and allows them to guarantee their client's professional project structures and organisation as well as precise cost-and-programme controlling without losing sight of the overall design aspiration. Their unique blend of challenging design and efficient project management finds its expression in an architecture which always begins with listening.

Waugh Thistleton Architects

Waugh Thistleton Architects are a London-based practice with an international reputation for leadership in the use of sustainable materials. The practice advocates sustainability in the widest sense, focusing not solely on energy in use, but on embodied energy and longevity. A world leader in engineered timber and a pioneer in the field of tall timber buildings Waugh Thistleton prioritise the principles of reduce, reuse, recycle in their work, producing innovative and imaginative design solutions that harness leading technologies and minimise their impact on the environment. Kieran Walker is an Associate at Waugh Thistleton and was the Project Architect for the Dalston Works project in East London.

01

Chapter One

Introduction

Introduction

Guy Marriage

DOI: 10.4324/9781003123873-1

Introduction

In 2020, our planet has well over 7 billion humans living on it and that number is projected to rise to up to 13 billion by 2050. Where will all those extra people live? While humans traditionally lived rural lives, more recently we have lived in suburban flat lands: wasteful in terms of land use and illogical in terms of transportation to and from places of work. This then is the crux of the matter: if we continue to grow, do we continue to build outwards into the suburban wastelands that surround our horizontal cities; or do we concentrate, build upwards, and live in urban apartment clusters? Is our experience with Covid-19 going to change the way we build and live? After all, it is certainly having a difference in the way that we work.

There are many factors that go into housing decisions – prime amongst them is not only the price of land, but also the availability of amenities, the cost of electricity or other sources of energy, the seismic condition of the land, the quality of the subsoil, the external cladding and weather-proofing of the building, the closeness to schooling, work and entertainment. Social and cultural amenities are more commonly available in cities than in suburbia and humankind desires to be with other humans: we are an intensely social species.

Indeed, with the population of *Homo sapiens* heading to a minimum of 10 billion by 2050, then at the very least, accommodation for an additional 3 billion people needs to be created over the next 30 years. There is some debate about whether the human population will continue to rise (Andregg, 2012; Liotta and Miskel, 2012), or perhaps level off (Livi-Bacci, 2017), or possibly even drop like a stone (Bricker and Ibbitson, 2019), but regardless, in the next 30 years, we will have to build housing like never before (Pagett and Smit, 2013). It is clear that for the continued existence of our prolific species, we must stop building on arable land, spreading out over green fields with a sea of low-level housing. The obvious consequence is that the future of our housing is likely to be more vertical and denser than in the past: the future of humankind is undoubtedly urban.

Urbanisation

Cities will continue to intensify regardless of health scares, with existing low-rise buildings increasingly replaced by medium-density or high-rise buildings. That's inevitable. Countries that once prided themselves for being rural idylls are increasingly admitting that actually, people like to live near other people and that in fact, urban life is growing at a considerable rate. Aotearoa New Zealand, once thought of by many as a green and peaceful land full of sheep, forests and mountains with a scattering of humans (and perhaps some hobbits), is now one of the more urbanised nations on planet Earth. In 2020, over 87% of the NZ population, now live in cities, although nowadays there's a lot fewer sheep. China, for centuries, consisting primarily as a nation of agricultural workers living in small rural villages, is urbanising rapidly as well, with over 50% of the greater Chinese population now living in cities. Over half a billion Chinese people are now urbanites, mostly living in concrete tower blocks. All over the world, cities are being built at great speed (often with not a lot of sophistication), to house their inhabitants as quickly as possible. Cities are the termite-mounds of human civilisation: a visual manifestation of the way we like to live (Wilson, 2020).

There's one big barrier to this argument however, and that is that, unlike termites, the way we are building now is not sustainable. Creating multi-storey communities from buildings made from concrete is neither a sensible nor responsible use of our increasingly scarce planetary resources. Sharp-edged construction-quality sand for use in both concrete and glass is running out (Torres et al., 2017). Concrete is an extremely energy-intensive product to create and use due to the immense amounts of CO_2 produced when cement is made. Concrete is largely a one-way street as a material: most demolition concrete goes straight to landfill. Steel is also extremely energy intensive to create, although it is excellent at recyclability. It is arguable that there is one material that we should be using more of in the future and that we can readily grow more of: big slabs of timber. Chapter 7, in particular, examines the relative benefits of timber over concrete and steel (Figure 1.1).

This book therefore is of two things: both a general guide to the creation of good spaces for living in tall buildings, and also a pathway to constructing these tall apartment buildings out of engineered timber. Principles for apartment living are the same the world over despite many varied climates and socio-ethnic groups: people simply want a good, safe, quiet, well-designed, warm, dry space in which to live. The creation of a community of souls in an active urban environment, living high above the ground next to people they may not know and may never meet is a skilled task and needs careful attention.

Modern construction systems are similar around the globe: no matter what the building is made from, gravity has the same effect on a structure wherever you live, while lateral forces like earthquakes, high winds and hurricanes do their best to rip it all apart. The book attempts to give an insight into the design aspects peculiar to high-rise living as well as a head-start on the detailing characteristics needed to cope with tall apartments, especially those made from modern engineered solutions like mass timber.

That's all assuming, of course, that the world will want to live in multi-storey apartment buildings in this post-Covid future. Some are predicting a push

Figure 1.1 Multi-storey mass timber residential building.
Source: Hayes, after Acton Ostry.

to return to the havens of suburbia, commuting into the city each day in single-occupant vehicles, much as we have for most of the last century (Wainwright, 2020). Many others widely presume that with the forced adaption of 'working from home' that we may never return to the cities at all. Are cities doomed? Are apartment buildings doomed? I certainly hope not.

Cities and disease

This coronavirus may be novel, but the effects are not new. Despite our inclination to live in cities, humans have always been susceptible to health risks when put in crowded conditions. Sanitation is the most important factor: without fresh water sources for drinking and bathing, and without adequate drainage systems for voiding sewerage, many cities have had waves of disease reduce their populations considerably.

The most destructive disease to hit cities in the last two millennia was the bubonic plague, causing death rates as high as 30% in some countries, arriving in waves between the 12th and 17th centuries. The plague (or black death) wiped out much popular inclination for urban living, as it spread rapidly via the vector of fleas living on rats, who enjoy living in cities as much as we do. People left the cities in droves and headed for the countryside – unfortunately taking the plague with them (Defoe, 1722). What is apparent in Defoe's *Journal of the Plague Year* is just how similar are people's reactions to spreading disease are, even today. Lockdown is not new. Self-confinement has many years of precedents, as has the effort to escape from confinement. Mask-wearing was also used to survive back then, particularly by the plague doctors. Vaccination dates from Jenner's cowpox investigations in 1796. Obsessive hand-washing is a relatively new one though.

Cholera epidemics (caused by sewerage contaminating supplies of drinking water) affected many cities around the world until 1854 when Doctor John Snow deduced the problem by tracing the source of contaminated drinking water to the Broad Street water pump in London's Soho and solved it by the simple act of removing the pump handle (Hempel, 2006). New York similarly suffered from waves of disease for many decades until a good fresh water source was well sourced (Glaeser, 2011).

More recently the 1918 influenza epidemic caused millions of deaths – more deaths than in the First World War that immediately preceded it. The 1918 flu was not caused or spread by cities, but like any highly contagious disease it took advantage of the close proximity of people, spreading rapidly in the troop ships that were returning soldiers back home.

Our current preoccupation with disease is the Covid-19 virus, which has caused considerable death and disruption around the world. Unlike previous waves of corona viruses like SARS, this had a more immediate and wide-spread effect. Our modern inclination for extensive tourism, mass gatherings and prolific air travel spread this air-borne disease immediately around the globe. On the other hand, our modern means of communication provided answers to this issue as well: immediate worldwide lockdown advice that was taken up by most world leaders. Not all populations believed the advice or followed it, but at the stage of writing, the first wave of combating the virus was largely over. The worst of pandemics was overcome in certain places around the world, essentially, by the simplest of simple measures: stay at home, wear a mask, and wash your hands. But by the time the second wave peaked in December 2020, many people had tired of waiting and refused to stay at home. Predictably, death rates soared again until the Covid vaccines were administered.

While cities such as New York and London have had massive death tolls, many other cities worldwide have not. Clearly, the mere fact of being in a city does not cause disease: Hong Kong, for instance, has a density far higher than that of New York, but has had a Covid-19 disease rate and death rate that is miniscule by comparison, powered by citizen vigilance and awareness of just how fragile the human condition truly is. The key appears to be preparedness to act: cities and countries that entered lockdown earlier rather than later have far lower death rates. Staying home, self-isolating, washing your hands and disinfecting door handles have entered into our *modus operandi* for continued good health, but many people could not stomach even those simple measures, preferring to gather *en masse* and party. Whether this can be blamed on the inherent human want to communicate with others, or cultural differences in our willingness to follow rules, or whether it is just a matter of stubborn nature and stupidity, the simple request to stay at home, which can save lives almost instantly, was ignored by many (Gelfand, 2020).

Diseases and apartment buildings

As the figures coming out of Hong Kong and Taiwan show, the mere act of living in an apartment building does not necessarily equate with the spread of disease. The exact vectors for transmission within a city are uncertain, but while close proximity is an issue, apartment living has solid, impermeable walls that keep residents perfectly safe. Hong Kong, despite having one of the highest densities of any city on the planet, had an impeccably low Covid-19 transmission rate throughout the first lockdown. This is due to the tightly organised population locking themselves down, and insisting on mask-wearing protocols along with mandatory hand-washing. Isolation is the key to success, and in this aspect, living in apartment buildings is as safe as living in suburban homes. As a primarily air-borne disease, Covid's means of transmission is more by breathing than by touching, but regardless, regular disinfecting of frequently-touched surfaces like door handles and elevator buttons is all part of the story.

Works cited

Andregg, Michael. 2012. *Seven Billion and Counting: The Crisis in Global Population Growth*. Minneapolis, MN: Twenty-First Century Books.

Bricker, Darrell and Ibbitson, John. 2019. *Empty Planet: the Shock of Global Population Decline*. London: Robinson.

Defoe, Daniel. 1722. *A Journal of the Plague Year – Being Observations or Memorials, of the Most Remarkable Occurrences, as Well Publick as Private, Which Happened in London during the Last Great Visitation in 1665*. London: Penguin Classics (2003).

Gelfand, Michele. 2020. "Why countries with 'loose', rule-breaking cultures have been hit harder by Covid." *The Guardian*, February 1, 2020. Available at: https://www.theguardian.com/world/commentisfree/2021/feb/01/loose-rule-breaking-culture-covid-deaths-societies-pandemic

Hempel, Sandra. 2006. *The Medical Detective: John Snow, Cholera, and the Mystery of the Broad Street Pump*. London: Granta Books.

Liotta, P. H. and Miskel, James. 2012. *The Real Population Bomb: Megacities, Global Security & the Map of the Future*. Dulles, VA: Potomac Books Inc.

Livi-Bacci, Massimo. 2017. *A Concise History of World Population* (4th Edition). London: Wiley Blackwell.

Pagett, Norman and Smit, Josephine. 2013. *The End of More: Infinite Demand on Finite Resources Is Making Humankind Unsustainable*. Davenport, IA: Lees Publishing.

Torres, Aurora, Liu, Jianguo, Brandt, Jodi, and Lear, Kristen. 2017. "The world is running out of sand," *Smithsonian Magazine*, September 8, 2017. Available at:https://www.smithsonianmag.com/science-nature/world-facing-global-sand-crisis-180964815/

Wainwright, Oliver. 2020. "Smart lifts, lonely workers, no towers or tourists: architecture after coronavirus." *The Guardian*, April 13. Available at:https://www.theguardian.com/artanddesign/2020/apr/13/smart-lifts-lonely-workers-no-towers-architecture-after-covid-19-coronavirus

Wilson, Ben. 2020. Metropolis: *A History of Humankind's Greatest Invention*. London: Jonathan Cape.

02

Chapter Two

Street life

Street life

Sara Shabahang and Guy Marriage

DOI: 10.4324/9781003123873-2

Introduction

Humans are a highly social species and like to live in company with each other. No other species likes to build as much as humans with only termite mounds rivalling us in the complexity of our collective habitat. Humankind has consistently lived in groups, such as walled compounds, villages, towns and now our present conurbations. David Frye puts it succinctly in the book Walls (Frye, 2018), that humankind shifted both physically and psychologically from being hunter-gatherers to a more sedate and well-behaved lifestyle once we started building walls around our collective living places. Walls keep wind, wolves and wandering wayfarers out, while providing a place of sanctuary within. Being urban meant we could become urbane. Walls lead to cities. Cities allow us to have more time for leisure, to be closer to the urban action (Wilson, 2020).

Village

Urban living has, for most of our existence on planet Earth, been based on the village. A small number of people, pooling resources to create ways of life that were not always revolving around where the next meal was coming from. Urban life is human life: Neolithic, Bronze Age, and Iron Age peoples helped hunter-gatherers evolve into hamlet, village and small town dwellers and as we progressed, our towns eventually grew into cities.

City

Cities are places of action and delight: of theatre and of congregation, of parliament and of mercantile trade. Little wonder then that the history of humanity is concurrent with the history of great cities and great battles for their control: it is only in the last 200 years that some people have begun to consider the suburb as the natural human habitat, particularly in the Anglo-American world. Suburbs, however, are not a true part of a city – they exist with a status that is neither city nor rural, but is simply <u>sub</u>urban: perhaps a lesser experience than urban. With suburbs comes an inevitable increase in the number of cars and sprawl.

Cities traditionally grow where there exists some natural fortification or barrier or trade advantage: an ocean anchorage, the side of a river, atop an island outcrop of rock, or the crossing of main routes by sea or by land. Petra, Jordan 'a rose-red city half as old as time' (Burgon, 1846) was occupied by the Nabataeans from around 7000 BCE alongside an ancient trading route and a waterway, although these days it is just a bare stone shell and a tourist trap. The oldest continually occupied city in the world is Luxor (ancient Egyptians knew it as Thebes), dating from 3200 BCE on the banks of the Nile. Damascus (circa 2000 BCE) and Jericho (1000 BCE) are also ancient urban locations, still occupied to this day. Living was not in apartment buildings though – more intermingled townhouses, with property boundaries indistinct or blurred. You can see this even today in small Greek hill-towns, where stairs to one dwelling are built over the roof structure of another dwelling below.

It is not for nothing that the Silk Road from China to Europe has spawned a hundred small cities along its route: cities thrive and grow on trade. Byzantium was at the crossroads between East and West; road and sea. If and when the trade route dried up, the city died. Cities such as Troy were built and rebuilt several times on the same site despite intermittent destruction: proof, if any was needed, that location is indeed the most important determinant in siting a city. The Bronze Age ended this first spurt of civilisation around 1177 BCE and, along with it, the knowledge of how to build tall buildings was lost (Cline, 2014).

Ancient empires

With ancient Greek and Roman civilisations (even the Romans referred to the Greeks as the 'ancients') we start to truly chart the rise of the city as an urban metropolis. Athens, the modern city of the Greeks, is built in, around and largely on top of the ruins of the ancient city, still clinging to the massive marble rock of the Acropolis. Rome still retains a great deal of structure from over 2,000 years ago. Both Athens and Rome were continuously embattled by other cultures wanting to besiege their city gates, and so by their very nature they were largely secured through the building of massive stone walls. Inhabitants lived within the walls of Rome in *insula*, the fore-runner of modern apartment blocks, sometimes several storeys tall. Tellingly however, *insula* is the word for island in Latin: a city block of housing that sits proudly above the street, independent of the waves of humanity washing below.

As always, the rich wanted to be at the centre of the action and so they scored the best locations in the city centres, despite the stench from human excrement in the streets below. As always, the poorer inhabitants had to move outwards, the start of a long history of suburban sprawl. The direction of urban growth was almost entirely upwards (with a bit of outwards), although it must be noted that in Cappadocia (as the area is known in modern-day Turkey), entire cities like Derinkuyu (near Ürgüp in Nevsehir) were carved out as many as five levels underground, to escape both marauding infidels and cold winter climates. This building system saves on façade construction costs but has a limited outlook for those that enjoy views out of their windows.

Medieval and Renaissance times

The fall of Rome in the 5th century AD (476 AD) meant the end of Western rule for the next millennium. Elsewhere empires kept growing in between spurts of warfare: China grew steadily but kept its populations largely rural and disconnected from the power in Beijing. Thalassocracies grew to govern the Mediterranean waterfront cities: sea-faring empires based on water-borne strength, with the Serenissima Repubblica di Venezia (Republic of Venice) ruling the seas for 1,100 years. In medieval Europe, living conditions were crowded and often squalid for the numerous poor, with most city dwellers living in crowded townhouses, whole families in one room. Constricted by the perimeter walls of old cities, there was little room to grow, except outside the city, unprotected. Occasionally fires would raze the old city, with the Great Fire of London (1666) being perhaps the most well-known, but the cities were usually rebuilt on the same lines as before.

While the Renaissance in Italy grew courtesy of powerful families like the Medici in city-states like Firenze (Florence), wars with neighbouring regions like Sienna always kept a firm cork on the bottle of unfettered growth. Housing for the Medici family and others grew into urban fortresses and palazzi: functioning as apartments but only for the extended family and their servants. Rooms opened off rooms, and circulation was usually around a central courtyard so that all comings and goings could be observed. While some people liked to live 'apart', most urban living was still centred around the family.

Urban design

The world now has well over 50% of the human population living in urban situations, and in some areas, over 80% living urban lives. We are, today, undoubtedly a very urban species, functioning to our best (and sometimes also worst) when living in cities and their surrounding suburbs. For the most part however, our cities have not been planned, they have simply grown, and grown, and grown. As our cities continue to expand and multiply, we can no longer rely on fortune and good luck in their planning. We now rely on a many-tentacled science known as urban design: despite the repetitive use of the word, there is no precise definition for it (Frey, 2003). Placed on a scale somewhere between urban planning and architectural design, urban design has been described as the design of public spaces or the practice of creating a single composition of many buildings (Moughtin, 2003). Urban planning is the science of environmental, transportation and land-use planning, and has a relationship with landscape architecture, policy recommendations, administration and urban design. Urban design examines the process, scale and product of our cities while also describing the different approaches: the political, social, physical, economic and environmental.

There are many ways of looking at the process of urban design, categorised neatly by Carmona et al. (2010). There are those who view urban design as a visual and artistic exercise, such as Cullen (1995) and Rasmussen (1964). Here, aesthetic points of view take preference, with streetscape defined as a series of picturesque viewpoints, in a more natural, organic, and certainly informal city planning than is capable from an all-pervasive grid. Other views on urban design are that the city should be created from the aspect of best social usage, where it is the people themselves that make the city, more than the buildings that enclose them. A third approach is one of physically moulding spaces and places to mix diverse activities and create specific identity and desired aesthetic aspects (Bentley, 1985; Jacobs and Appleyard, 1987; Lynch, 1984).

The philosophical point of view has always been very important in urban design and architecture (Alexander, 1977; Broadbent, 2003; Venturi et al., 1977). Historicism, modernism, new modernism, post-modernism and many diverse contemporary styles have all had very different approaches to shaping cities. For instance, modernist urban design used a non-historic approach, promoting healthy practical spaces that were highly vehicle-oriented (Middleton, 1983). The complicated nature of urban design needs specialists from different disciplines, including planners, architects, urban designers, landscape architects and traffic planners (Moughtin, 2003).

The procedure for designing our future cities is crucial for sustainable development: sustainable cities are needed as a sensible response to global warming (Carmona et al., 2010). Sustainable development in urban areas needs to include the three main aspects of the social, environmental and economic, sometimes known as the pillars of sustainability (Biddulph, 2007; Carmona et al., 2010). Sustainable urban forms were categorised by Jabareen (2006) into four models: (1) compact city, (2) eco-city, (3) neo-traditional development (new urbanism) and (4) urban containment. The first three are considered to be the most important in regard to sustainability. Moreover, to create a sustainable urban form, density, compactness, diversity and mixed land use should be considered in parallel with advanced design and management systems.

Street pattern

Modern cities and transportation are inseparable subjects which affect city layout: street patterns shape urban neighbourhoods and define accessibility. Historically, urban streets not only provided connectivity between different destinations but they also provided spaces for vital social gatherings such as 'meeting, trading, hawking, busking, bear-baiting, public speaking and pillorying' (Marshall, 2004). While there has always been a connection between important places in cities and the liveability of the streets, the urban revolution truly happened with the invention of the automobile. Modernism requested other transportation infrastructures like motorways and highways: streets as places for people became roads as routes for cars. The historic pattern of the streets with a concentration on places has changed to a pattern of distribution.

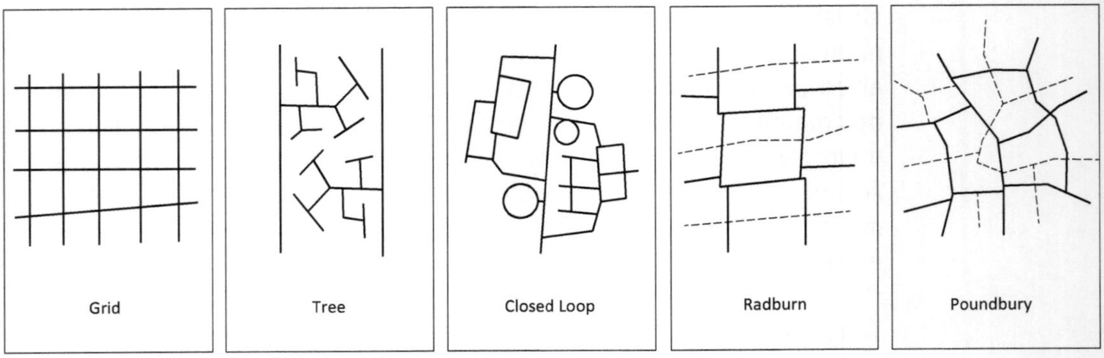

Figure 2.1 Road layout patterns.
Source: Hayes/Marriage, after Biddulf (2007).

Streets and roads are classified based on function, material, shape, size, geometry, traffic volume and hierarchy of distributions. Any street has multiple characteristics, designed under different disciplines by both urban designers and civil engineers. Street type, street pattern and street hierarchy are the common terminologies for any of above classification. However, these terms sometimes create confusions and there is no distinct difference between them. In general, road hierarchy is based on capacity or traffic flow (i.e. arterial road, primary road, motorway, etc.), while street type classifies roads based on their functions (pedestrian street, local access street, shopping boulevard, cycle route, bus route, etc.).

Although street pattern and urban pattern are connected concepts, they are not the same things. Urban pattern refers to the structure of the settlement and street pattern refers to the shape of the network movement. The most famous typology of settlement defined by Lynch (1984) classified city forms as star, satellite, linear, rectangular grid, other grid (parallel, triangular, hexagonal), baroque axial network, lacework, inward city (medieval Islamic), nested city or current imagination (bubble, floating, underground). Modern street patterns are a lot simpler by contrast, and Biddulph categorised these into just five principle forms: the grid, the tree, the closed loop, the Radburn and the Poundbury. See Figure 2.1.

Urban mobility

The urban mobility paradigm can be summarised into different steps. The distribution of land uses is determined by the location of human activities, while the infrastructure provided for human activities is determined by the transport system. The transport system distribution further determines accessibility and finally, accessibility can affect land use locations.

Transport is a key land use. Cities provide a huge amount of land for transportation and parking. The transportation sectors are blamed for producing a third of CO_2 emissions (Yang and French, 2013). This has a terrible effect not only on the environment but also on human wellbeing. Making cities accessible for different users by different transportation means is a key factor in urban design and city planning. Pedestrian-friendly and bike-friendly approaches in cities attract a huge amount of attention because of health, environmental and economic benefits. Many researchers point out that high population density, mixed-used high-rise buildings, as well as good public transportation, provide good conditions for walkable cities (Agrawal and Schimek, 2007; Cervero and Kockelman, 1997). People who live in high-rise buildings spend less money on transport than those who live in single-detached houses (Mattson, 2020). Population density has a negative effect on energy consumption (Newman and Kenworthy, 1989) and CO_2 emission (Susilo and Stead, 2009). Dense environments also affect the taste of residents choosing a vehicle: they prefer smaller cars (Bhat and Sen, 2006). Job density and street network connectivity affect Vehicle Miles Travelled (VMT) in a positive manner (Ding et al., 2017), while well-developed public transportation has a negative effect on CO_2 emissions per capita (Yang et al., 2015).

Low-density single-house development increases car dependency for residents by increasing distances between destinations. It also results in low walkability in areas. Because low density urban areas are usually far from city centres, they demand more time on transportation. Conversely, an increase in density results in reduced VMT (Litman and Steele, 2017). Urban densification can happen by building towers and building blocks: although the construction of towers demands high investment, it offers better returns.

Urban planners have proposed Transit-Oriented Development (TOD) (Calthorpe et al., 2001) and Greenspace-Oriented Development (GOD) principles (Bolleter and Ramalho, 2019). TOD means increasing density around public transportation locations and corridors; GOD means considering sufficient public green space when densifying around public transportation.

Green space

Scarcity of green space is a primary concern of many urban neighbourhoods. The benefits of green space have been extensively discussed (Maas et al., 2006; Wolch et al., 2014), although land use in terms of the balance between density and distribution of green space is a matter of debate between researchers (Ng et al, 2012). Green space distribution in cities generally follows a hierarchical pattern (Biddulph, 2007), with people travelling longer distances to large green areas and local green areas being much smaller in size. The area of green space, the scale of the urban area and the accessible distance for the individual are thus fundamental to green space distribution.

Kleerekoper et al. (2012) highlight the cooling role of green space in cities as it leads to evaporation, vegetation to provide shading, and heat transfer at night. Moreover, their study states the main types of green space in cities are urban forests (parks), street trees, private green areas in gardens, and green roofs and façades. Swanwick et al. (2003) also distinguish between different types of urban green and grey space.

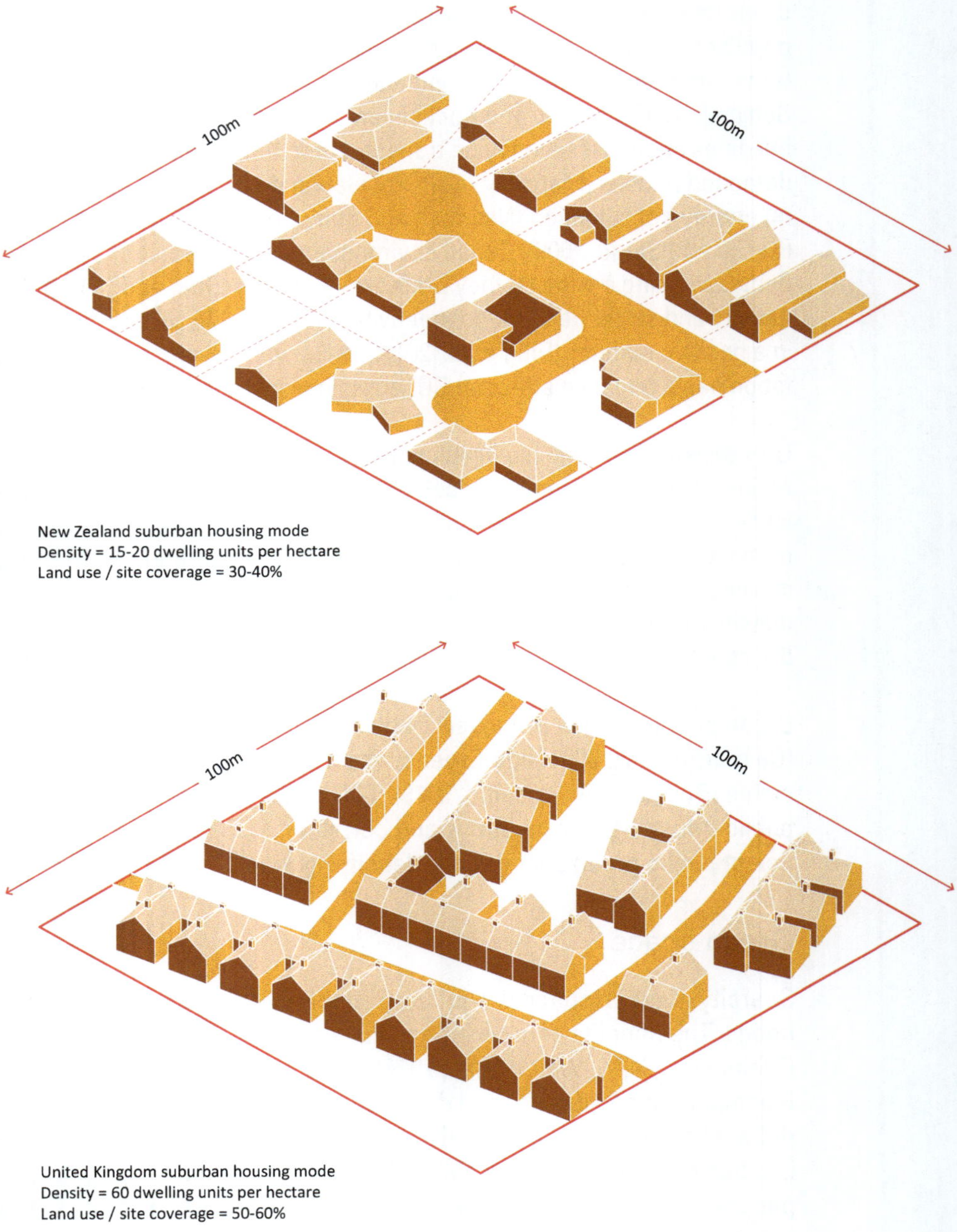

Figure 2.2 Suburban density versus terrace housing.
Source: Hayes/Marriage, after Cheng (2010).

Density

The meaning of the term density varies widely. Originally used in physics as a quantity of mass within a unit volume, in urban studies it has been defined as the quantity of people or housing units per area. These certain units could be footprints, building heights or number of building storeys, while the unit area could be site coverage, neighbourhood or region. Densities can be measured (such as floor area ratio or dwelling density) or qualitative. Qualitative density cannot be measured but can be readily perceived and understood. Perceived density is the focus of environmental psychology and the analysis of perceptions of crowding. In other words, it is a combination of qualitative and quantitative densities.

Density can be measured as gross or net. When all areas are included in an area calculation it is called gross density. When roads or green space are excluded from a calculation, it is called net density. There are many different ways of measuring density, some of which are listed below in Table 2.1. For example, external density is the number of people measured at the neighbourhood scale while internal density is people per room or floor area. Cheng (2010) divided the concept of physical density into four main groupings based on population, buildings, geography and social matters.

Commonly however, when speaking of density today, we typically simplify matters of density down to just three simple measures: low density, medium density, and high density (Environment, 2011). We do not have one common set of standards or terms for these concepts, nor standard measures (Table 2.2). Even within a single country, different cities may have their own definitions.

Urban density

Dense-built environments are associated not only with health and safety problems but also with sustainability. Density is usually used as a means to control and quantify built environment characteristics. Among the different concepts of density, density of buildings and population are correlated to each other (Dovey and Pafka, 2014). Jacobs believed that a high-density, low-rise neighbourhood with 80% site coverage leads to urban diversity (Jacobs, 2016). There is no relationship between mental health and the density of urban areas at the neighbourhood scale (Cohen and Sherrod, 1978), although the United Nations has called three persons living in the same room a slum (UN-Habitat, 2004).

Table 2.1 Possible density measures

People density	Regional density	Ratio of a population to the land area of a region
	Residential density	Ratio of a population to the residential land area of a region
	Occupancy density	Ratio of the number of occupants to the floor area of an individual habitable unit
Building density	Plot ratio (floor area ratio)	Ratio of the total gross floor area of a development to its site area
	Site coverage	Ratio of the building footprint area to its site area
	Dwelling density	Number of houses to the given area
Urban density	Network density	Measure of network interconnectivity that is largely synonymous with spatial permeability
	Site/parcel density	Dwelling units or residential population per site/parcel area
	Block density	Dwelling units or residential population per area of a city or neighbourhood block
	Net neighbourhood density	Number of dwelling units or residents divided by the area of the neighbourhood, excluding city-wide uses such as parks or large commercial areas
	Net neighbourhood residential density	Number of people or residential dwelling units divided by the total residential land area
	Gross neighbourhood density	Number of dwelling units or residents divided by the total area of the Neighbourhood – no exclusions are made
	City or municipal density	Number of dwelling units or residents divided by the total area of the city or municipality

Source: Definitions from Forsyth (2003, pp. 5–6).

Table 2.2 Some definitions of density by country

	Low density	**Medium density**	**High density**
UK (source: TCPA, 2003)	Less than 20 du/ha	30–40 du/ha	60+ du/ha
UK (source: Levitt and McCafferty, 2018)	35–90	90–250	250–350
USA	25–40	40–60	110+
Israel (source: Churchman, 1999)	20–40	–	290+
New Zealand (source: Department of Internal Affairs, 2019)	10–25	50–100	125–250 du/ha

Sources: As noted above.

That said, high density is correlated with sustainability as it leads to less transport energy consumption. Ewing and Hamidi (2017) in their research on the quality of life in suburbia argued that transportation, personal health and economy are better in dense areas (Table 2.3).

Table 2.3 Compact city versus urban sprawl

	Positive	**Negative**
Compact city	- decreases travel distance - increases public transportation and decreases commuting costs - increases street life vitality - increases walkability - better public transport - higher public safety - higher job opportunities - more social interaction	- increases travel time because of traffic jams in cities - increases traffic congestion - open space and green space limitation - less sunlight - overcrowding - increase crime - light pollution - noise pollution
Urban sprawl	- increases space per person - lowers house prices - more green space	- increases travel distance between two destinations - increases reliance on private cars and consequently higher CO_2 emissions - destruction of farms and wildlife - additional use of energy - decreases work opportunities because of low population and high travel distance - increases commuting cost - causes health issues such as obesity because of using cars more and walking less - infrastructure issues such as higher costs for drainage systems and garbage disposal - gives rise to residential segregation

Source: Ismael (2020).

A compact city policy is usually an attempt to define the urban form and try to stop unplanned urban growth. Multifunctional land use is part of the debate for enhancing urban life by saving space, creating safer environments and being more economically advantageous (Rodenburg et al., 2003). Both the compact city and multifunctional land use could be achieved by using a higher urban density design approach. Higher urban density is seen as a sustainable approach in smart growth theories by many researchers, because it leads to less land consumption as well as lower energy consumption. However, there is not a total agreement about measurement methods. For example, both areas in Figure 2.2 have a density of 75 dwelling per hectare while having different urban forms. This weakness in the relationship between residential population and density of population has been pointed out by some researchers (Berghauser Pont and Haupt, 2009). They suggest measuring urban footprint per resident in a given area instead.

Urban sprawl

Population growth expands the cities. It is predicted that by 2030, the occupied area of human population will have grown by another 1.2 million km^2 of land worldwide (Seto et al., 2012). Sprawl city is one type of urban form and is usually considered the opposite of a compact city. The existence of urban sprawl dates back to the appearance of early cities, when people decided to escape from busy crowded cities and settle on the land around them. In modern times, the term 'sprawl' was first used by the British government in 1935 and by the American government in 1937. In general, sprawl can refer to a low density built area regarding building size, demographic factors and more space per person. Cities may sprawl based on political, cultural and religious reasons: sometimes based on deliberate policies, at other times despite of them.

Planners have agreed on the interrelationship of land use and transportation in cities. Deficiency in providing enough options for transport such as walking and bicycling is one of the main problems of suburbia (Atash, 1994). In other words, the expansion of cities means increasing the distance travelled by people and an increase in car use, with sprawl considered as a dynamic or static phenomenon – or both. From the urban design point of view, urban sprawl is discontinuous development (Peiser, 1989) due to high land consumption and urban infrastructure demands, with side effects such as destroying farmlands and wildlife habitat (Figures 2.3, 2.4 and 2.5).

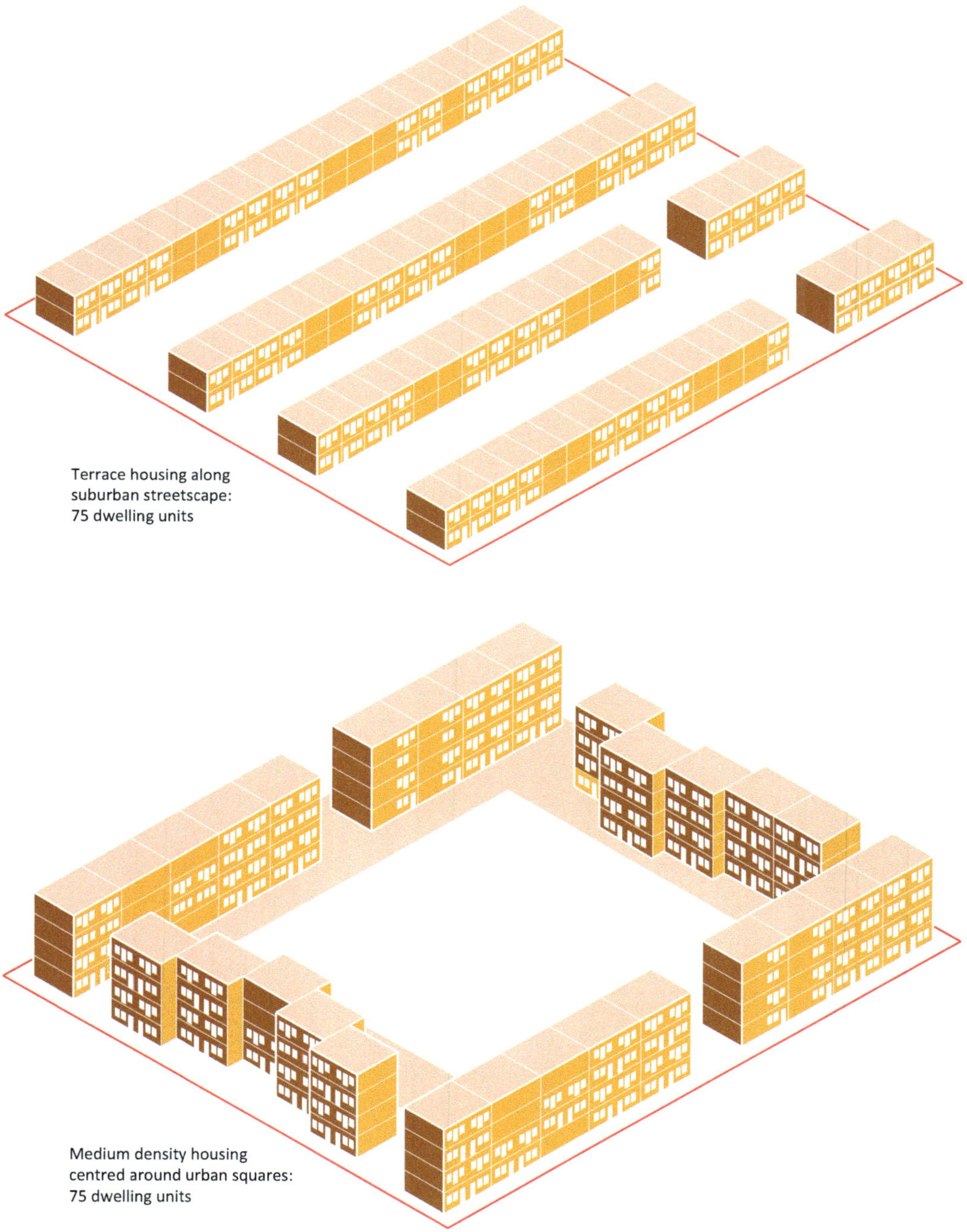

Figure 2.3 Row housing versus medium density.
Source: Hayes/Marriage after Berghauser (2009).

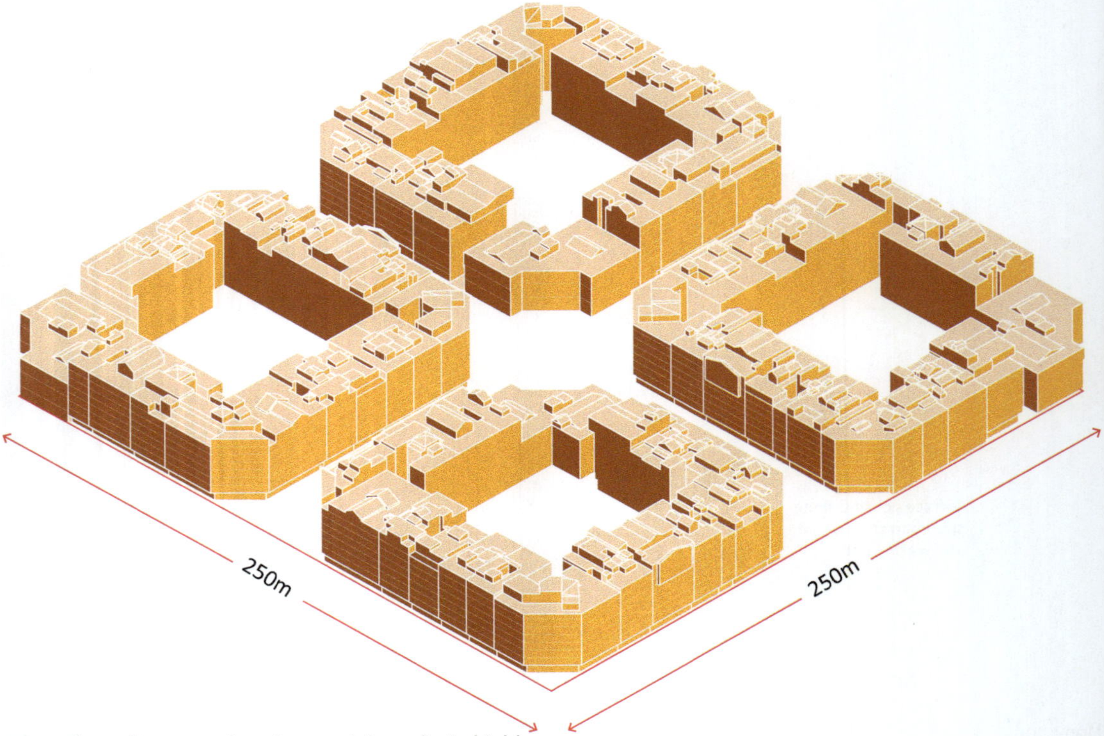

Barcelona: European housing model results in highly successful urban life

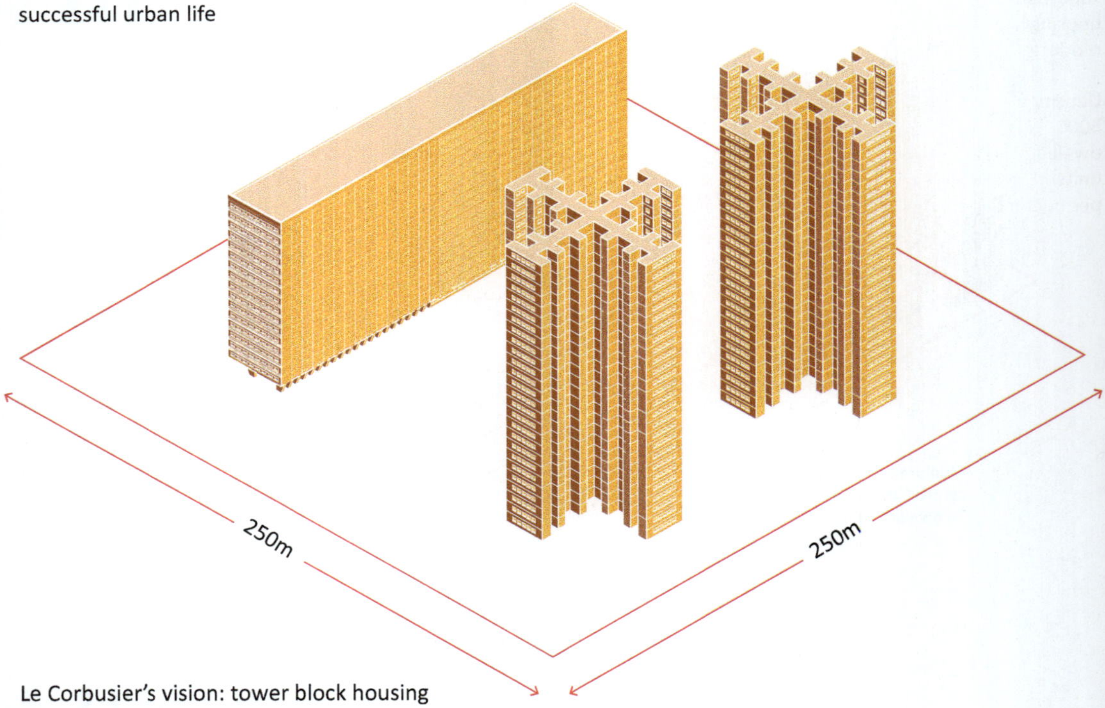

Le Corbusier's vision: tower block housing results in alienation and loss of urbanity

Figure 2.4 Barcelona density versus Corbusian density.
Source: Hayes/Marriage.

Hong Kong
suburban
housing
mode

Density =
3000
dwelling
units
per hectare

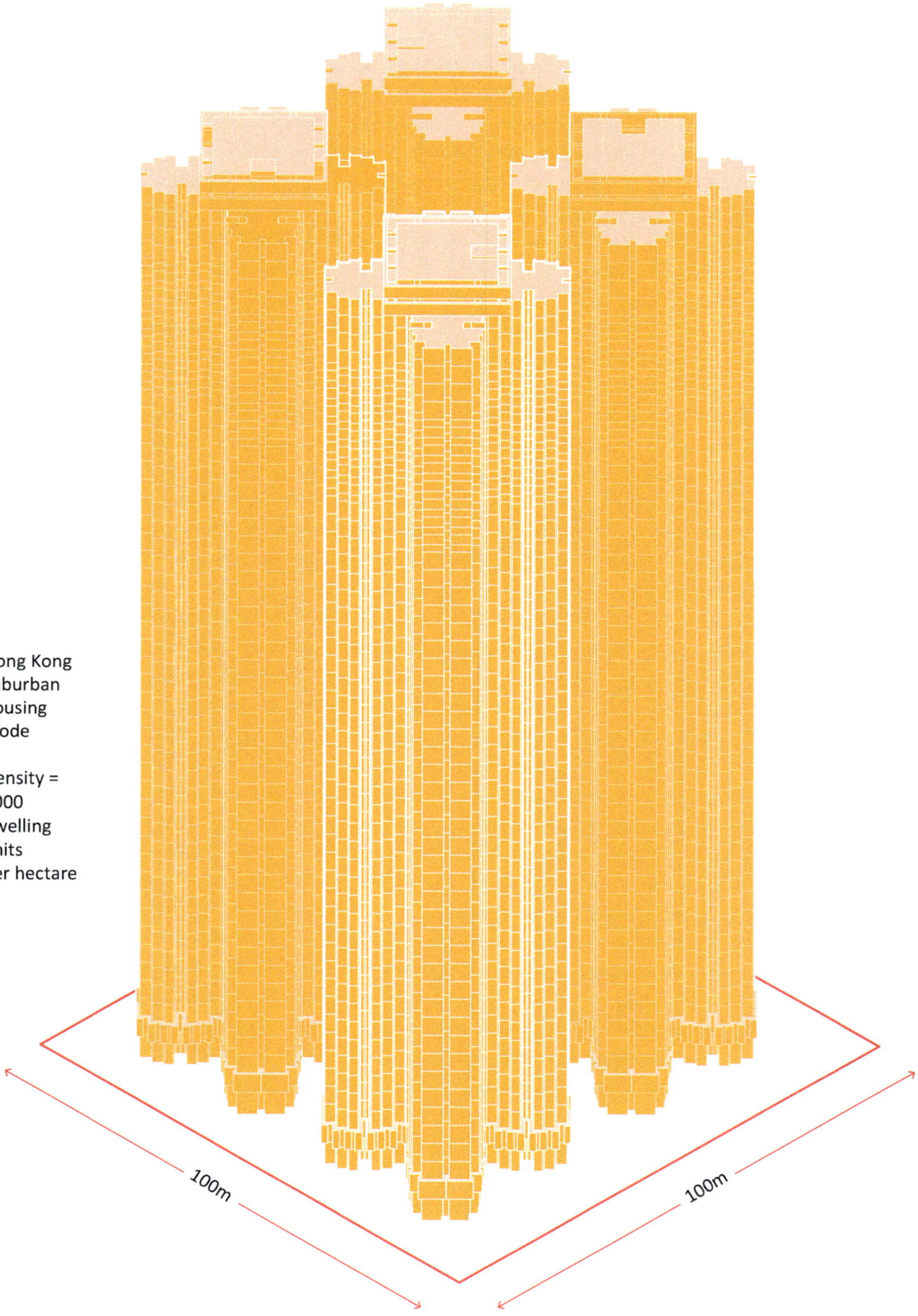

Figure 2.5 Hong Kong density – New Harmony cruciform tower.
Source: Hayes/Marriage, after Hong Kong (1994).

Works cited

Agrawal, A.W. and Schimek, P. 2007. Extent and correlates of walking in the USA. *Transportation Research Part D: Transport and Environment*, 12(8), 548–563.

Alexander, C. 1977. *A Pattern Language: Towns, Buildings, Construction*. Oxford: Oxford University Press.

Atash, F. 1994. Redesigning suburbia for walking and transit: emerging concepts. *Journal of Urban Planning and Development*, 120(1), 48–57.

Bentley, I. 1985. *Responsive Environments: A Manual for Designers*. London: Routledge.

Berghauser Pont, M.Y. and Haupt, P.A. 2009. Space, density and urban form. NAI Publishers.

Bhat, C.R. and Sen, S. 2006. Household vehicle type holdings and usage: an application of the multiple discrete-continuous extreme value (MDCEV) model. *Transportation Research Part B: Methodological, 40*(1), 35–53.

Biddulph, M. 2007. *Introduction to Residential Layout*. London: Routledge.

Bolleter, J. and Ramalho, C.E. 2019. *Greenspace-Oriented Development: Reconciling Urban Density and Nature in Suburban Cities*. Springer.

Broadbent, G. (2003). *Emerging Concepts in Urban Space Design*. London: Taylor & Francis.

Burgon, John. 1846. *Petra, a Poem: To Which a Few Short Poems Are Now Added* (2nd Edition). Oxford: MacPherson.

Calthorpe, P., Fulton, W., and Fulton, W.B. 2001. *The Regional City*. Island Press.

Carmona, M., Heath, T., Tiesdell, S., and Oc, T. 2010. *Public Places, Urban Spaces: The Dimensions of Urban Design*. London: Routledge.

Cervero, R. and Kockelman, K. 1997. Travel demand and the 3Ds: density, diversity, and design. *Transportation Research Part D: Transport and Environment, 2*(3), 199–219.

Cheng, V. 2010. Understanding density and high density. In Ng, Edward (ed). *Designing high-density cities, 37–51*. London: Routledge.

Churchman, A. 1999. "Disentangling the Concept of Density." *Journal of Planning Literature, 13*(4), 389–411.

Cline, Eric. 2014. *1177 BC: The Year Civilization Collapsed*. Princeton, NJ: Princeton University Press.

Cohen, S. and Sherrod, D.R. 1978. "When density matters: environmental control as a determinant of crowding effects in laboratory and residential settings." *Journal of Population, 1*(3), 189–202.

Cullen, G. 1995. *The Concise Townscape.* London: Routledge.

Department of Internal Affairs. 2019. *Glossary.* Available at: https://www.dia.govt.nz/diawebsite.nsf/wpg_URL/Resource-material-Building-Sustainable-Urban-Communities-Glossary?OpenDocument

Ding, C., Liu, C., Zhang, Y., Yang, J., and Wang, Y. 2017. "Investigating the impacts of built environment on vehicle miles traveled and energy consumption: differences between commuting and non-commuting trips." *Cities, 68,* 25–36.

Dovey, K. and Pafka, E. 2014. "The urban density assemblage: modelling multiple measures." *Urban Design International, 19*(1), 66–76.

Environment, M.f. 2011. *Medium-Density Housing in New Zealand.* Available at: https://www.mfe.govt.nz/more/towns-and-cities/medium-density-housing

Ewing, R. and Hamidi, S. 2017. *Costs of Sprawl.* London: Taylor & Francis.

Forsyth, A. (2003), Measuring Density: Working Definitions for Residential Density and Building Intensity, Design Center for American Urban Landscape Design Brief 8, Minneapolis, MN, DCAUL. http://annforsyth.net/wp-content/uploads/2018/05/db9.pdf

Frey, H. 2003. *Designing the City: Towards a More Sustainable Urban Form.* London: Taylor & Francis.

Frye, David. 2018. *Walls: A History of Civilisation in Blood and Brick.* London: Faber & Faber.

Hong Kong Government of Planning Department. 1994. *Hong Kong Planning Standards and Guidelines.*

Ismael, H.M. 2020. "Urban form study: the sprawling city—review of methods of studying urban sprawl." *Geo Journal,* 1–12.

Jabareen, Y.R. 2006. "Sustainable urban forms: their typologies, models, and concepts." *Journal of Planning Education and Research, 26*(1), 38–52.

Jacobs, A. and Appleyard, D. 1987. "Toward an urban design manifesto." *Journal of the American Planning Association, 53*(1), 112–120.

Jacobs, J. 2016. *The Death and Life of Great American Cities.* Vintage.

Kleerekoper, L., van Esch, M., and Salcedo, T.B. 2012. "How to make a city climate-proof, addressing the urban heat island effect." *Resources, Conservation and Recycling, 64,* 30–38.

Levitt, David and McCafferty, Jo. 2018. *The Housing Design Handbook: A Guide to Good Practice*. London: Routledge.

Litman, T. and Steele, R. 2017. *Land Use Impacts on Transport*. Victoria Transport Policy Institute, Canada.

Lynch, K. 1984. *Good City* Form. MIT Press.

Maas, J., Verheij, R.A., Groenewegen, P.P., De Vries, S., and Spreeuwenberg, P. 2006. "Green space, urbanity, and health: how strong is the relation?" *Journal of Epidemiology & Community Health, 60*(7), 587–592.

Marshall, S. 2004. *Streets and Patterns*. London: Routledge.

Mattson, J. 2020. "Relationships between density, transit, and household expenditures in small urban areas." *Transportation Research Interdisciplinary Perspectives*, 8, 100260.

Middleton, R. 1983. "I. The Use and Abuse of Tradition in Architecture." *Journal of the Royal Society of Arts*, 131(5328), 729–739.

Moughtin, C. 2003. *Urban Design: Street and Square*. London: Routledge.

Newman, P.W. and Kenworthy, J.R. 1989. "Gasoline consumption and cities: a comparison of US cities with a global survey." *Journal of the American Planning Association*, 55(1), 24–37.

Ng, E., Chen, L., Wang, Y., and Yuan, C. 2012. "A study on the cooling effects of greening in a high-density city: an experience from Hong Kong." *Building and environment*, 47, 256–271.

Peiser, R.B. 1989. "Density and urban sprawl." *Land Economics*, 65(3), 193–204.

Rasmussen, S.E. 1964. *Experiencing Architecture* (Vol. 2). Cambridge, MA: MIT Press.

Rodenburg, C., Vreeker, R., and Nijkamp, P. 2003. "Multifunctional land use: an economic perspective." In *The Economics of Multifunctional Land Use*, 3–15. Shaker.

Seto, K.C., Güneralp, B., and Hutyra, L.R. 2012. "Global forecasts of urban expansion to 2030 and direct impacts on biodiversity and carbon pools." *Proceedings of the National Academy of Sciences*, 109(40), 16083–16088.

Susilo, Y.O. and Stead, D. 2009. "Individual carbon dioxide emissions and potential for reduction in the Netherlands and the United Kingdom." *Transportation Research Record*, 2139(1), 142–152.

Swanwick, C., Dunnett, N., and Woolley, H. 2003. "Nature, role and value of green space in towns and cities: an overview." *Built Environment (1978–)*, 94–106.

UN-Habitat. 2004. "The challenge of slums: global report on human settlements 2003." Management of Environmental Quality: *An International Journal,* 15(3), 337–338.

Venturi, R., Stierli, M., and Brownlee, D.B. 1977. *Complexity and Contradiction in Architecture (Vol. 1): The Museum of Modern Art.*

TCPA – 2003. (Town and Country Planning Association). *TCPA Policy Statement: Residential Densities.* London: TCPA.

Wilson, B. 2020. *Metropolis: A History of Humankind's Greatest Invention.* London: Jonathan Cape.

Wolch, J.R., Byrne, J., and Newell, J.P. 2014. "Urban green space, public health, and environmental justice: the challenge of making cities 'just green enough'." Landscape and *Urban Planning,* 125, 234–244.

Yang, J. and French, S. 2013. "The travel—obesity connection: discerning the impacts of commuting trips with the perspective of individual energy expenditure and time use." *Environment and Planning B: Planning and Design,* 40(4), 617–629.

Yang, W., Li, T., and Cao, X. 2015. "Examining the impacts of socio-economic factors, urban form and transportation development on CO_2 emissions from transportation in China: a panel data analysis of China's provinces." *Habitat International, 49,* 212–220.

03

Chapter Three

Looking back

Looking back

Emina Kristina Petrović and Guy Marriage

DOI: 10.4324/9781003123873-3

Introduction

Throughout the history of humankind and around the world, it is possible to observe a range of different housing forms, often with intricate and complex underlying stories (Shoenauer, 2000). Within this complex web, two threads stand out as the most clearly defined: detached individual houses and 'houses' grouped together into larger buildings, such as apartments, flats or tenements. This book investigates the latter, and this chapter looks into the forms of the apartments in those parts of the world where apartments are commonly used.

It is important to emphasise that any discussion of housing inevitably closely relates to how people see themselves and their own place in the social structure. On that level, consideration of different housing models closely relates to the social perceptions and aspirations at the time. For the same reason, in different parts of the world certain housing models can be seen as more or less desirable. The passion with which people discuss their individual suburban houses and urban apartments is clearly expressive of their social and cultural forces, and it has been argued that such preferences are not predominantly influenced by rational evaluations of actual qualities of the housing, but rather their complex social and cultural values and shared aspirations (Petrović, 2003). Consequently, this chapter focuses on the examples of preferences for urban apartment or similar living, and provides some insights into the social and cultural complexities underpinning these preferences.

The birth of the apartment

In order to track the reasons for modern appreciations for apartment living, it is useful to review the ideas associated with the emergence of the apartments during the second half of the 19th century, around the times of industrialisation and urbanisation. At the time, apartment buildings introduced a higher level of comfort with a group of rooms called an apartment, while retaining an urban monumental presence.

Within the European context, the medieval walled cities corresponded with the prevalence of reasonably dense and tall townhouses. Paris, with its successive expanding of the city walls, presents one excellent example of this tradition, and the Parisian *maison à allée* captures the transition from the medieval townhouses to the 19th century bourgeois apartment buildings. Similarly, while the word *appartamento* originated in the 16th century in Italy (literally 'living apart' from others; Middleton, 1992, 32), it is in Paris that *appartement* became more commonly used to describe a group of rooms used by one person or a family within a larger complex, gradually leading to becoming a synonym for a way of urban living (Figure 3.1).

Figure 3.1 French medieval housing.
Source: Marriage.

Maison à allée typically developed haphazardly. In contrast to the medieval townhouses, which were normally inhabited by the owner as an individual house, *maison à allée* were subdivided and individual rooms or groups of rooms were separately rented out (Loyer, 1988, 22, 24). Often rooms used by one family were distributed over several floors, in different parts of the building, which at times required people to pass through the rooms of other tenants (Marcus, 1999, 21). The most desirable living spaces were closer to the ground and closer to the front of the building, while the building as a whole was shared by both a range of activities, including shops, workshops and housing, and different social classes (Loyer, 1988, 22, 24, 48).

Greater clarity of organisation and grouping of rooms developed in the *hôtel particulier*. Initially, the *hôtel* was a temporary urban residence of the French feudal aristocracy, a reduced urban replacement of their country chateau (Zuskin, 1999, 16). Internally, the rooms of the *hôtel* were organised into groups of four or five, forming an apartment for specific purposes, for the use of a particular person or for representative or social functions (Middleton, 1992, 32). By the 18th century, *hôtels* represented a system of various apartments organised within an externally unified and monumental whole. When the strict divisions between aristocracy and the bourgeois (middle class) were upended by the French Revolution (1789–1799), the *hôtels* were used as a basis for the subsequent bourgeois apartment housing.

The bourgeois apartment buildings of 19th-century Paris organised the previous jumble of rooms within the *maison à allée* into a clear and practical system. One apartment occupied one floor (later, larger floor areas were sometimes divided into two or three apartments), with one access from a public staircase and another from a service staircase leading to the kitchen and the service area of the apartment (Gallet, 1972, 65, 80). Servants had to climb to the top of the building each night to sleep. The main entry into an apartment was through an antechamber leading to the dining room, or through a *dègagement* to a salon, which provided a privacy filter which was previously unknown in Paris (Lipstadt, 1977, 34–47; Olsen, 1986, 119). In addition, these buildings offered a monumental entrance and overall façade, while retaining the presence of commerce on the ground and mezzanine levels. Yet, the new model enabled the separation of work and living, and more privacy, thus providing a solution to the Parisian desire for a marriage of the private home and urban public monumentality (Marcus, 1999). This allowed for the development of a desirable housing model for the French bourgeoisie, and as a consequence urban living was preferred by many, with the city periphery predominantly left to the poor and those less fortunate.

And lastly, with the invention of the *ascenseur* (lift or elevator) from 1840 onwards (and in common use by 1880), the social order of the apartment building was upended too. No longer were servants at the top: the most desirable apartments were now in the clean fresh air at the top of the building (Sutcliffe, 1981) (Figure 3.2).

Germany and Europe

In some other parts of Europe similar general evolution from the medieval townhouse of the walled towns to new models of reasonably higher density living was also taking place. Generally, innovation accelerated as urbanisation accelerated, and 19th-century developments for housing the growing bourgeoisie established many of the key trends for what followed.

Many Germanic regions had developed in a similar manner to France. As with the *hôtel particulier*, urban living was favoured by the higher classes and already in the 16th century, the Habsburg imperial court issued orders which forced the local population to accept court and state officials as tenants (Blau, 1999; Olsen, 1986). *Zinsplast*, rent-palace, was the purpose-built response, which included monumental, luxurious forms and reception spaces. In these residences, the focus was on the public side which meant that there were no specialised bedrooms, but rather sleeping

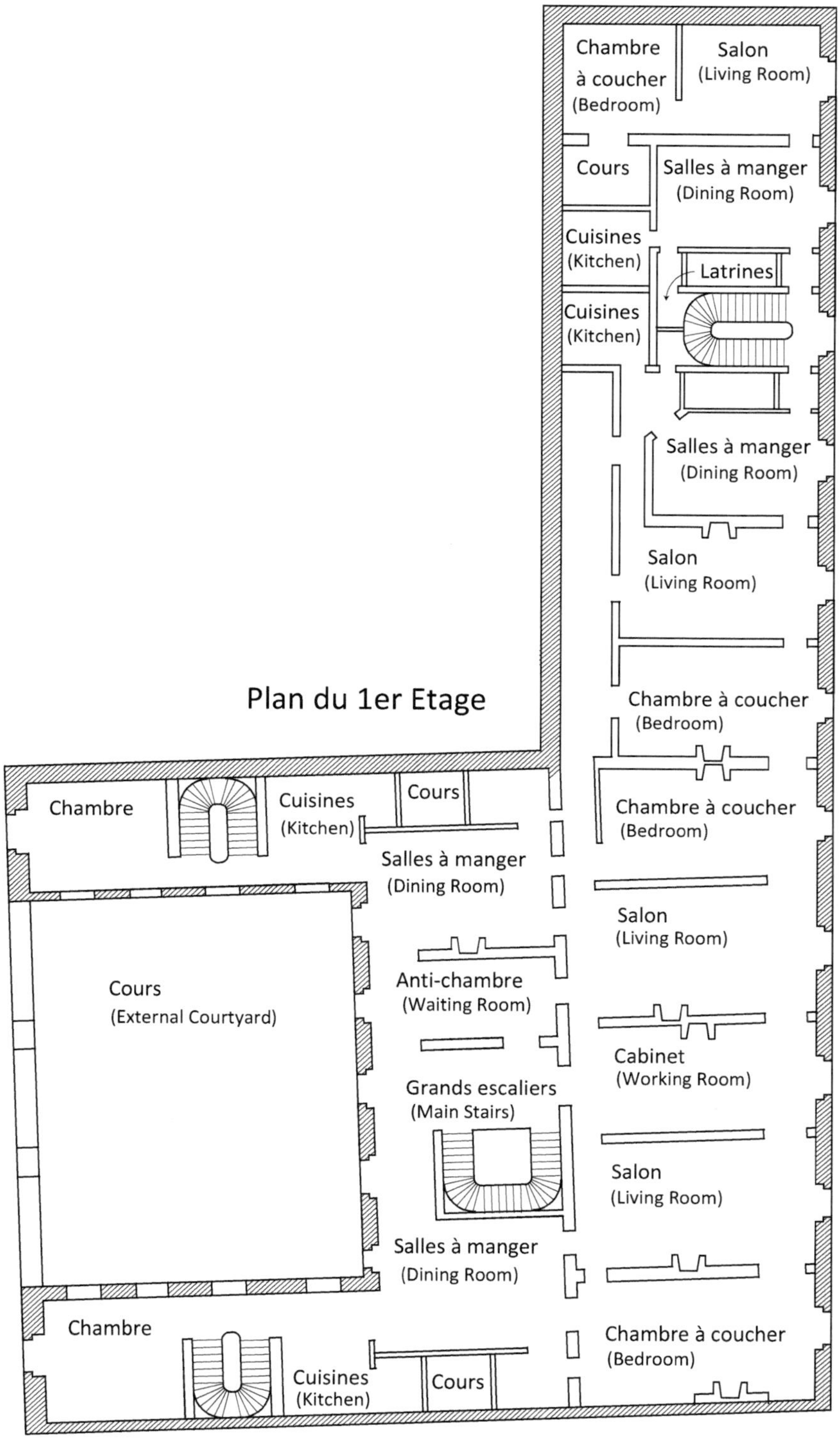

Figure 3.2 Floor plan of an apartment building on the boulevard St-Denis, Paris (1837).
Source: Marriage, after Marcus, *Apartment Stories*.

was improvised in some of the grand reception spaces (Olsen, 1986). Hallways were mainly absent, and the internal circulation was through the interconnected rooms. Nevertheless, in the Germanic regions, even at the peaks of urbanisation during the 19th century, centres of the larger cities remained commonly used for housing the highest classes (Olsen, 1986) (Figure 3.3).

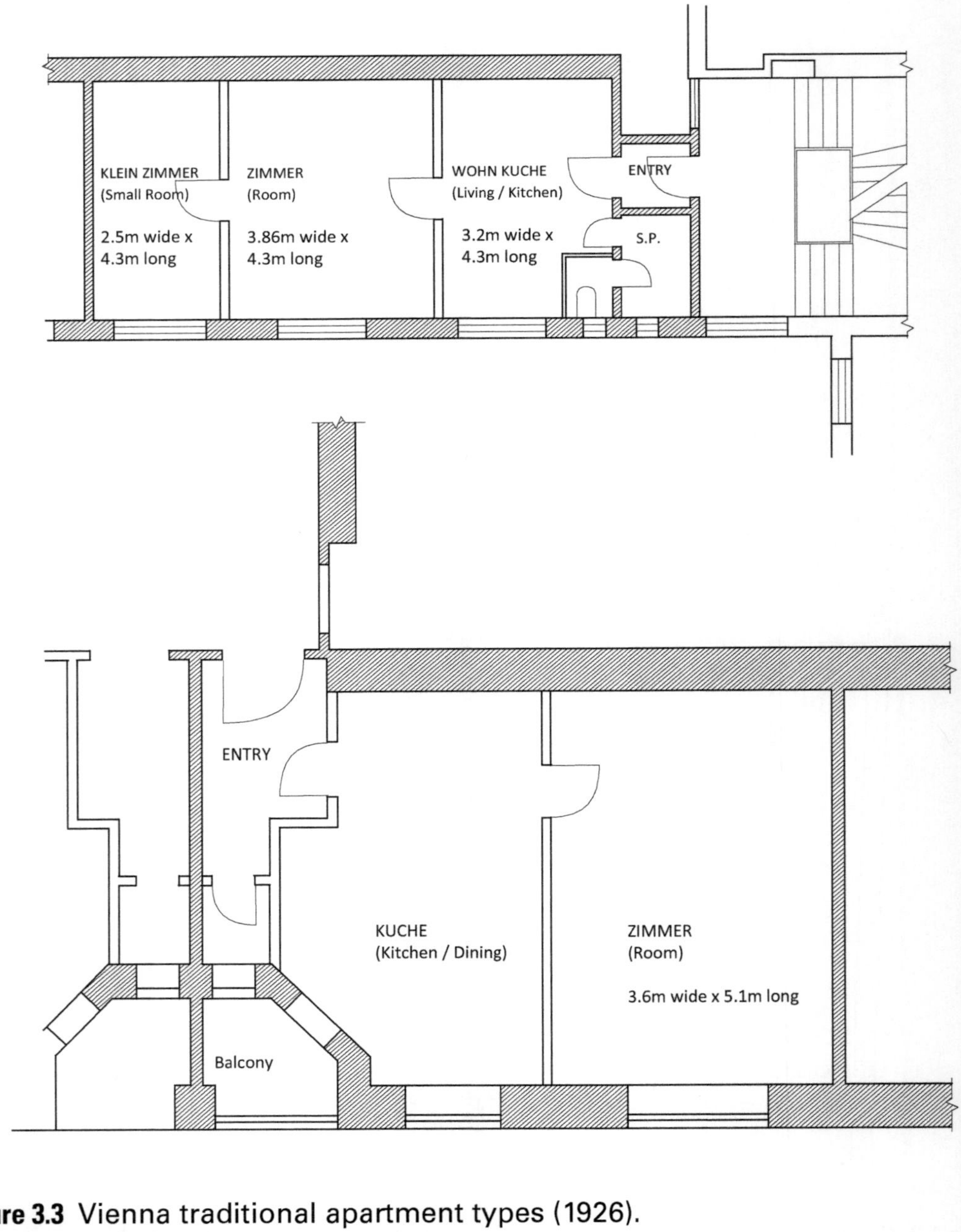

Figure 3.3 Vienna traditional apartment types (1926).
Top ($48m^2$), bottom ($38m^2$) – not to same scale.
Source: Marriage, after Blau, *The Architecture of Red Vienna*, p. 182.

In terms of form, *zinsplast* were organised around an open-air courtyard which was a common form throughout the region and known as a *hof-haus*. The *hof-haus* was derived from rural farm or vintners' houses of wine-producing districts, often with working and living spaces (Blau, 1999). These were buildings of two or more storeys and were seen as acceptable accommodation for anybody from the aristocracy, in *zinsplast*, to the poor in more reduced forms. With the pressure on land, this process intensified and from the 1840s, tall blocks organised around one or more courtyards became increasingly common (Sutcliffe, 1981). When providing small single room dwellings, these large complexes were called *mietskaserne*, rental barracks.

As with the *maison à allée* in Paris, the *zinsplast, hof-hauses* and the *mietskaserne* maintained the continuous street fronts, and little structure in the internal organisation of spaces: halls and rooms with specialised functions were mainly absent (Olsen, 1986). Echoing developments in other parts of Europe, especially in England and Paris, at the turn of the 20th century, the specialisation of rooms and organisation into apartments became more common. Still, even at the peak of state-stimulated construction of housing during the Weimar Republic, it was accepted that providing a larger number of smaller, cheaper units was more favourable than attempting to construct good-sized units, which were still too expensive for many (Heynen, 1999). This in itself reinforced the normality of multi-unit dwellings in these regions, and maintained many elements of the monumentality achieved with the large façades and shared staircases.

During the 1920s and 1930s, German modern architectural developments leaned strongly towards functionalism and the past issues with internal organisation were resolved into rational, functional groupings of specialised rooms which varied in size and were grouped into clear zones (Petrović, 2003). Much of this development sets the scene for contemporary apartment designs.

Many other parts of Europe experienced similar trends with a range of multi-unit housing forms developing and being improved over time in response to the strong preference for urban living for all classes. Sophisticated culturally-appropriate design responses developed, and in many cases, the wealthier classes kept their strong presence in the urban centres, while the poorer were pushed out to the outskirts (Figure 3.4).

England

London's apartment journey was very different to that of Paris or Berlin. Despite being the centre of the British Empire and the world's first city to

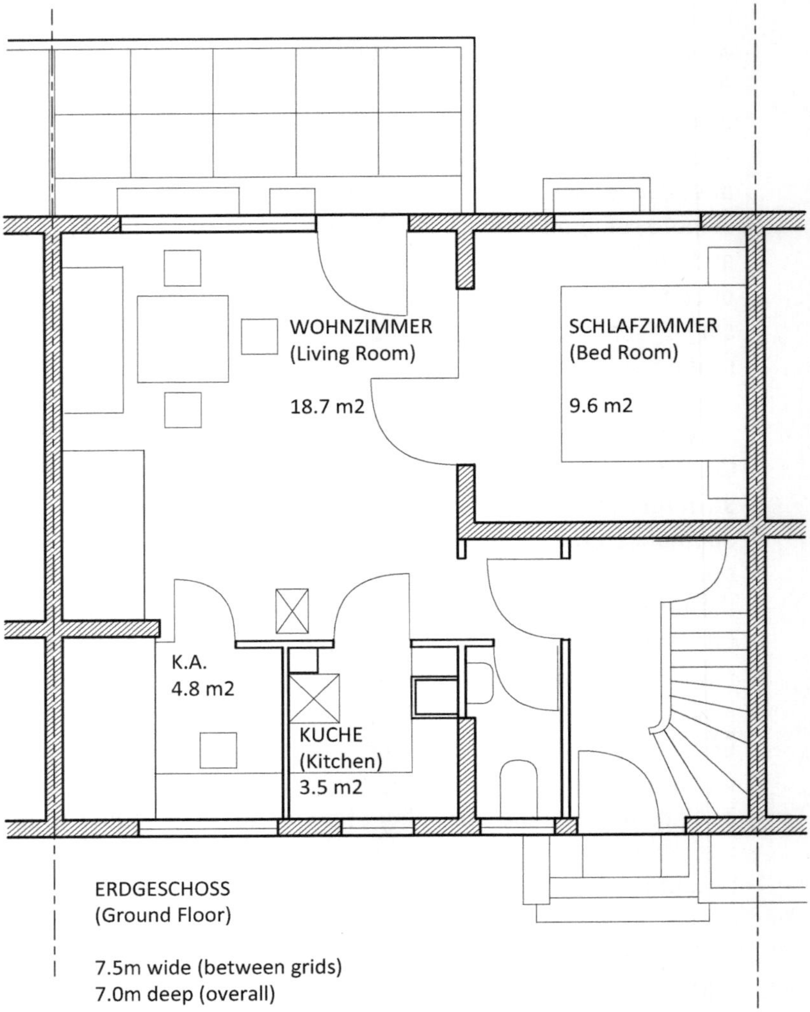

Figure 3.4 Frankfurt apartment in Siedlung Wasthausen.
Source: Marriage, after Ernst May, in Lane, *Architecture and Politics in Germany*, p. 102.

surpass 1million residents by 1811, London did not grow upwards as much as it expanded outwards into opulent suburbs, leaving the city centre for those too poor to move. London remained proud of its quirky medieval street pattern in the old city. London never rebuilt in the style of Paris: there was no Baron Hausmann to rework the city streets, despite Wren's efforts after the Great Fire. The cities of Westminster, London, Southwark continued to grow on the banks of the Thames – the boundaries are still hugely evident in the modern metropolis if you know where to look: Bishopsgate perhaps being the most obvious example of extreme wealth on one side of the street (the upmarket office blocks of the City of London)

Figure 3.5 Good example of functional housing for frictionless living.
Source: Marriage, after Klein, *Functional housing for frictionless living* (1928) in Bauer, *Modern Housing*.

and extreme poverty on the other side (run-down social housing in the East End borough of Tower Hamlets) (Figure 3.5).

As the city boundaries grew and merged, property ownership by the wealthy land-owners enabled large areas of their estates to be developed into a series of Squares, some with private gardens at the centre. A series of townhouses surrounded each square, later divided up into ever smaller and smaller flats. Although it has become apartment living over time, it was never designed as such, and similarly much British urban housing suffers from all the traits of trying to divide up a building into multiple ownership. Some purpose-designed apartment buildings in the 20th century, known as mansion blocks, were large, almost monumentally-proportioned buildings which were planned from the outset as apartments, with a grand entry point and imposing façade. Largely however, Britain has only really entertained good purpose-built apartment design in the last few decades, although with spatial standards that are amongst the lowest in Europe.

United States of America

In the USA, the cities in the eastern states were more likely to follow the preference for urban apartments as with many continental European cities, while individual suburban houses (similar to the English preference) dominated in many other parts of the USA. New York, founded by the Dutch in the 16th century, has undergone wave after wave of rebuilding in a manner similar to major cities in continental Europe. By 1975, over 98% of Manhattan families lived in apartments (Alpern, 1975). The Manhattan culture of luxury apartments, which included penthouse, duplex, and triplex apartments, set a new standard for the desirability of urban living and Manhattan even held the world record for the tallest buildings (including the Empire State Building and the Twin Towers). Apartment growth has been particularly prevalent in a city environment that includes a vibrant urban mix of commercial, retail and residential. Geologically, New York's upward growth has also been aided by the rock beneath the pavement: solid granite is a good base for building the foundations of skyscrapers.

The technological advancements developed for the construction of office skyscrapers were also applied to apartment buildings, often using very similar internal organisational approaches, with clear hallways, and structural systems often using brick and steel (Alpern, 1975; Leslie, 2013). For the deep plot sizes in Manhattan, the issue of access to views and light remained, and while this was better resolved in the luxury apartments (including offering a view of Central Park), servants and more modest apartments had limited access to these aspects. The Chelsea, built in 1883 as cooperative apartments (architect: Hubert, Pirsson and Co.) on Manhattan's 23rd St, was converted to hotel status by 1905 and remains highly sought after to this day. Original features like rooms for Maids may have moved on, and apartments have been subdivided many times since, but a quality of space within lingers on: high ceilings and character never go out of style (Figure 3.6).

Nevertheless, the suburban individual house model continued to be popular in most parts of the USA, and by the start of the 21st century the USA had a reasonably high homeownership rate of about 66%, with about 80% of that being detached individual houses (Lasner, 2012). Still, apartments remained more prominent especially in places like New York, which had a stronger history of apartment preference.

Since the 1950s, with the mass adoption of the automobile as the chief transportation method of North Americans, traditionally-formed

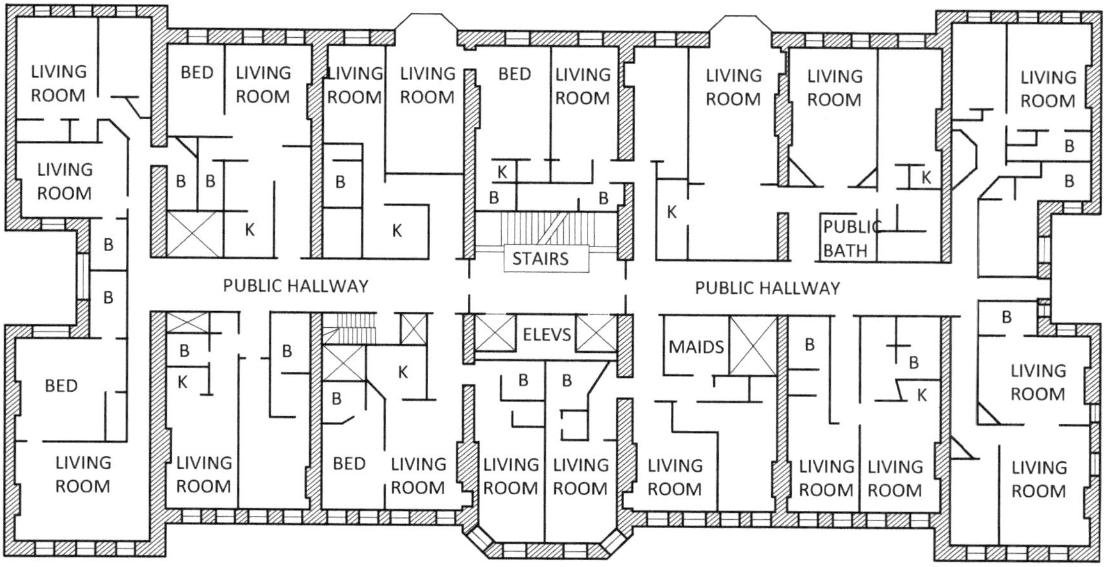

Figure 3.6 Hotel Chelsea.
Source: Marriage, after Alpern.

cities such as New York and Boston have been challenged by more vehicular-based cities such as Houston and Los Angeles (Glaeser, 2011). Not only do Houston and automobiles go hand-in-hand, Houston is largely unliveable without an air-conditioned car, an air-conditioned house and an air-conditioned place of work. Land is cheap, oil is cheap, and Houston sprawls low-cost suburbs to the horizon in every direction, while its freeways grow ever wider. Houston, long-term, is an unsustainable model.

Asia

It's impossible to summarise over a billion people in a few glib sentences, but needless-to-say, China is now at the forefront of urban living in terms of quantity, if not yet in quality. In particular, Hong Kong, Shanghai and Singapore are dense with apartments. While Hong Kong started off as nothing more than a fishing village a little over a 100 years ago, its extreme growth over the last 50 years has seen the inhabitants take the lead in apartment living. Because of the small land area of Hong Kong Island and Kowloon, accompanied by strict government land ownership, land has always been an extraordinarily expensive item. Spurred on by slums in the 1950s, the Hong Kong Housing Association has pioneered apartment living in a variety of forms: slab blocks, tower blocks, cruciform blocks and more, all featuring tightly-packed designs. The story of Asian apartments deserves a book all of its own (Figure 3.7).

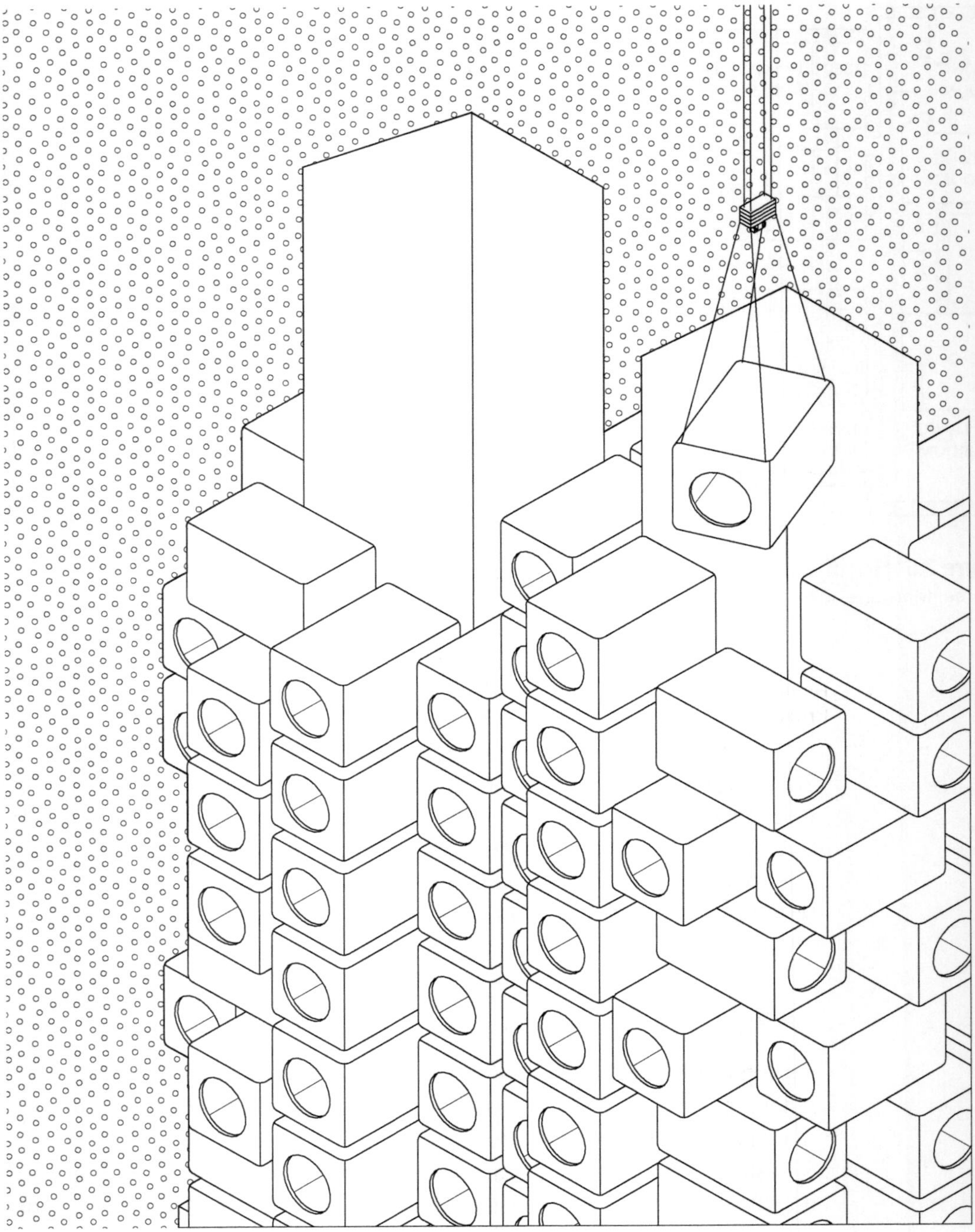

Figure 3.7 Nagakin Capsule Tower (architect: Kisho Kurokawa).
Source: Hayes.

Works cited

Alpern, Andrew. 1975. *New York's Fabulous Luxury Apartments: With Original Floor Plans from the Dakota, River House, Olympic Tower and Other Great Buildings.* New York: McGraw-Hill.

Blau, Eve. 1999. *The Architecture of Red Vienna 1919–1934.* Cambridge, MA: MIT Press.

Gallet, Michel. 1972. *Paris Domestic Architecture of the 18th Century.* London: Barrie & Jenkins.

Glaeser, Edward. 2011. *Triumph of the City: How Urban Spaces Make Us Human.* London: MacMillan.

Heynen, Hilde. 1999. *Architecture and Modernity: A Critique.* Cambridge, MA: MIT Press.

Lane, Barbara Millar. 2006. *Housing and Dwelling: Perspectives on Modern Domestic Architecture.* London: Routledge.

Lasner, Matthew Gordon. 2012. *High Life: Condo Living in the Suburban Century.* New Haven, CT and London: Yale University Press.

Leslie, Thomas. 2013. *Chicago Skyscrapers (1871–1934).* Urbana, Chicago, and Springfield: University of Illinois Press.

Lipstadt, Helene. 1977. "Housing the Bourgeoisie: Cesar Daly and the ideal home." *Oppositions,* No. 8 (Spring 1977): 41.

Loyer, Francois. 1988. *Paris Nineteenth Century: Architecture and Urbanism.* New York: Abbeville Press.

Marcus, Sharon. 1999. *Apartment Stories: City and Home in Nineteenth Century Paris and London,* 19. Berkeley: University of California Press.

Middleton, Robin. 1992. "Introduction." In *The Genius of Architecture; or, the Analogy of that Art with Our Sensations,* edited by Nicolas Le Camus de Mezieres, 32. Santa Monica, CA: The Getty Centre for the History of Art and the Humanities.

Olsen, Donald. 1986. *The City as a Work of Art: London, Paris, Vienna.* London: Yale University Press.

Petrović, Emina. 2003. "Your new house – or would a flat suit you better? Development of housing preferences in England, Paris, Germany and New Zealand." Master's thesis, Victoria University of Wellington.

Shoenauer, Norbert. 2000. *6,000 Years of Housing*. London: WW North and Co.

Sutcliffe, Anthony. 1981. *Towards the Planned City: Germany, Britain, the United States and France, 1780–1914*, 116. Oxford: Basil Blackwell.

Zuskin, Rochelle. 1999. *The Place Vendome: Architecture and Social Mobility in Eighteenth-Century Paris, 65*. Cambridge: Cambridge University Press.

04

Chapter Four

Apartment block typologies

Apartment block typologies

Guy Marriage

DOI: 10.4324/9781003123873-4

Introduction

We've got Le Corbusier to thank (or curse) for the uptake of modern apartment buildings. When Corbusier laid down his guidelines for living in the 'Radiant City', he envisaged people residing in the sky, sweeping away of the tired old European examples of crowded brick terraced housing. Children would play in rooftop gardens, windows would be wide, horizontal and independent of the structure behind, while the ground floor would be a place for people to walk freely under buildings, with the apartments themselves sailing majestically overhead on *piloti* dotted around a green natural landscape. Cars and trains would whizz quietly past on nearby aerial highways, while planes would criss-cross the sky in an orgy of transport delight. Life would be modern, magnificent and gloriously uncrowded (Cohen and Benton, 2008).

The truth, as it happens, has turned out quite different from Corb's prophecy. World cities tried to implement his vision but most of his dream turned out to be an urban nightmare. Buildings hovering effortlessly above the ground may look good in pictures but in reality they just accumulate dirt, dust and druggies in their undercroft spaces. Cars and trains whizzing noisily past apartment windows are universally loathed by those within. Parents did not enjoy having their children playing in the sky and preferred them to play at ground level, directly under strict parental supervision. Corbusian apartment schemes like the earnestly modernist Pruitt-Igoe apartments in St Louis, Missouri (architect: Minoru Yamasaki), were completed by 1954 but proved to have serious design faults that meant the 11-storey slabs started to look run down by 1957 (Bristol, 1991). They were ignominiously demolished on 16th March 1972; Charles Jencks famously dated the 'Death of Modernism' to this moment as the explosives sank the Corbusian dream. JG Ballard's wickedly sardonic novel *High-Rise* (1975) savagely lampoons the experience of urban living in a tall tower block, noting the severe effects of societal and physical breakdown (Table 4.1). All designers of modern apartment blocks need to understand the lessons from schemes like Pruitt-Igoe, Ballard's *High-Rise* and Ronan Point (Heathcote, 2020). So, what have we learnt?

Access to the apartment block

The lessons start at the ground floor. Oscar Newman strongly argued for the concept of defensible space: the ability to control the semi-public space outside your door (Newman, 1972). Having the public walking and driving underneath your building does not achieve this feeling of safety: having living spaces directly adjacent to the public at ground floor means passing pedestrians can look directly inside your house, with no white picket fence in the way. To integrate the development truly into the neighbourhood, a ground floor of retail or commercial spaces

Table 4.1 Typology of residential buildings

Typology of building	Key feature
Stand-alone housing	- smaller lot sizes - not attached to other dwellings but close to neighbouring buildings - two or three storeys in height - can be part of a larger master-planned development.
Semi-detached or duplex dwellings	- two side-by-side dwellings contained within one building - one dwelling is usually the mirror image of its partner - two or three storeys in height.
Terrace or row housing	- row of identical or very similar attached dwellings that are joined on one or both sides to other houses - the 'end terrace' house can be different to the rest of the terrace - sometimes can be joined by garages between houses and can either be built into the terrace and accessed from the front or can be accessed by a rear laneway - two or three storeys in height. - staircase can be longitudinal or traversal - terrace houses can sit back to back
Town house	- apartments are usually single storey self-contained units within a larger building, but sometimes apartments have more than one storey - usually, there is common access to a core stairwell - private open space is a courtyard or garden on the ground floor or on balconies on upper floors - often rubbish storage is communal and post boxes are in one central place
Courtyard houses	- floor plan type can be L-shaped or patio - courtyards are book-ended by building walls or neighbour walls - it is linked to other houses - one to three floors
High-rise building	- smaller footprint - shared access - more than seven floors - the important difference between high-rise and medium-rise is the requirement for special engineering because of height
Skyscrapers	- feature buildings of the 20th century - very cost-effective in areas with high land value and high population

Source: Ministry for Environment (2011).

means that urban street life can proceed peacefully at ground level, while upper floors can be accessed only by the urban residents. Mixed-use developments are therefore highly recommended – perhaps harder to achieve under the planning regulations but more successful long term for residents.

Block design

Block layout is the first decision, planning how the building (or buildings) will function in the urban environment. Consider orientation to sunlight, to prevailing winds, and to neighbouring buildings (more in Chapter 5). If you can, set back the façade slightly at the entry, to allow a transition space between the public footpath and the private property of the interior. Allow for trees or garden space near the entry to lighten the effect on the streetscape. Even better, integrate open communal garden space within your building as well.

Assuming an urban context, access is likely to be directly off the street. In a suburban site, primary access for residents will more likely be from a car parking facility nearby, or in the basement. Consider where pedestrians from the building will want to walk: where are their likely destinations and walking routes? In the city, access is likely to be directly off the street (residents will walk, bike, or taxi to their destination) or from a basement car park, while in a suburban site primary access is more likely from a surrounding at grade car park.

Plan for a readily accessible loading bay for removal trucks and delivery vehicles to load and unload. Plan for this to be as close and convenient to the circulation core as possible, but avoid having it directly outside the front door. A discrete separate rear entrance for rubbish collection is desirable.

Ground floor entry

The entry to the building is important, and security here is vital. It is both the most heavily used and visually prominent part of the interior. Make it work seamlessly and make it of quality materials. Clear and simple access is the key. Unlike a suburban residence, an apartment building rarely has a back door – it is all about the front door. Lock this access point securely, as no one wants unknown persons hanging around the main entry doors.

Ground floor lobby

Once inside the front door there needs to be a ground floor lobby. The space can be used to check your snail mail (if such a thing survives into the future), sign for courier packages or for pizza delivery workers to gather their wits and loose change. In our apartment building, we have a comfortable leather-clad padded seat for friends to wait, which often gets used as a mini book shop, where recently-read books are left out for others to enjoy. The ground floor entry lobby is the welcome lobby to the whole building – make the most of it. Directly off the lobby should be access to the core (more on cores in Chapter 10). Some of the uses that you will need to cater for in the ground floor lobby are:

- Mailboxes – one per flat – and somewhere for parcels as well.
- Comfy seat – for more than one person – a space to wait.
- Informal meet and greet area – to pick up pizza or discuss your pet dachshund.
- Access for public utilities metering, etc. – water, electricity, phone and data, gas, sprinklers, and, most importantly, the Fire Brigade who will connect their equipment in there in time of emergency.
- A mirror – to check you still look stunning before setting forth into the public realm.

Apartment access: corridors

Vertical circulation cores will lead to the chief horizontal circulation, usually in the form of an internal corridor, or a continuous external 'gallery' walkway spanning across several apartments. This creates a public side for access – anybody on that floor can directly access the front door. An external walkway, while it may look good on paper, is unpleasant to live on as a tenant. Decades of negative experience with single aspect, gallery access apartment buildings bear out this lesson. Avoid, if possible (Newman, 1996).

An internal corridor is standard for hotels and can work well for small apartment blocks with studio apartments or one-bedroom apartments, but it creates problems for larger apartment blocks. If the corridor has apartments off to one side only, it is a single-loaded corridor. If there are apartments off both sides, then it is a double-loaded corridor.

Make corridors short and sweet, or better still, design them out altogether. Long thin corridors are uncomfortable and alienating for residents to walk through, and they are a costly waste of space. Avoid bends in long corridors for security reasons and ensure clear view shafts where possible from the lift/stair lobby to the front door of the apartment.

Broad, wide-open corridors feel much nicer but result in more expensive apartments (due to wasted space in non-lettable areas). If you do have to have a corridor, it should always be a minimum of 1.2m (4′) wide and this width should increase around individual apartment entries and overall entry lobbies. Good corridor lighting is also very important for the sanity and safety of the residents – natural daylighting is better, if at all possible.

Remember that your building will be used by children at some stage – maybe not often or maybe all the time – and despite Corbusier wanting to put the children on the roof, children will end up in the corridor, close to their caregivers. Make it nice for them and not annoying for the neighbours! If you don't want corridors at all, then you will need to design a tower building, with a central core, or a slab block with multiple cores.

Common areas: lobbies and corridors

The key to any commercial building success relies on a good net to gross ratio, which means that keeping the proportion of rentable/sellable space high while keeping the common area space down to a minimum. Core spaces in multi-storey apartment buildings are much smaller than in multi-storey commercial office buildings for a number of reasons: fewer lifts (elevators), no central toilets required, no central HVAC required, fewer mechanical services all round. There will always be a need for lifts (preferably always two) and at least two staircases (some jurisdictions will permit only one).

The length of corridors should be kept down to a minimum or abolished altogether – this will depend on your building plan shape and site size. Instead, if possible, design in an area where a community can form and bond over common ground – an atrium perhaps, a communal courtyard, a sky garden, or a roof deck.

Planning

Apartment layouts can come in endless variations around the world, but the following guiding principles will help you get the most from your site layout wherever you are:

Single aspect

When apartments face outwards in only one direction, they are termed 'single aspect'. This orientation naturally limits the amount of sunlight and natural ventilation making its way inside. It is crucial that the living area should face outwards, allowing the maximum amount of daylight into the apartment. Bedrooms require daylight and fresh air as well, so the width to depth ratio of the apartments is important: a wider apartment is preferable, but more costly. More crucially, bedrooms require privacy, which restricts where the bedrooms can go – usually at the edge of the apartment, with living space in the middle. Single-aspect apartments on either side of a central corridor will sentence apartment on one side of the building to remain in shadow while the other side basks in sunshine (Figure 4.1).

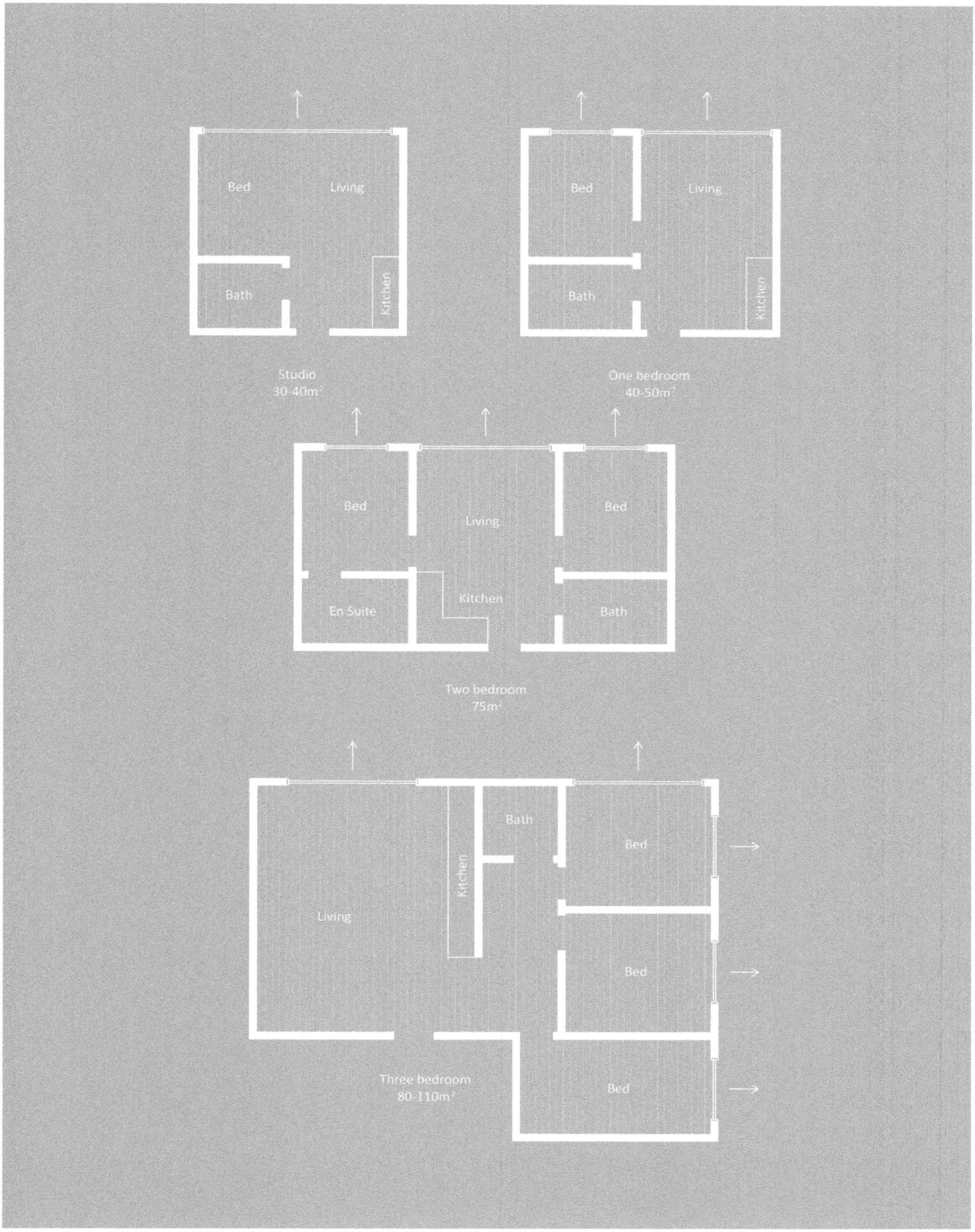

Figure 4.1 Single-aspect and dual-aspect apartments.
Source: Hayes/Marriage.

Dual aspect

Dual-aspect apartments have many advantages over a single-aspect apartment. Having two street frontages enables privacy to be designed into both ends of the apartment, allows opportunities for fresh air to cross-ventilate the apartment and offers possibilities of sunlight to two separate sides of the apartment. It allows bedrooms to be at either end or at both ends, while also allowing for a living room space that could span the entire block, giving light and space aplenty. The biggest issue with a dual-aspect apartment on a single level is that the access point limits where it can be situated. Single-level dual-aspect apartments usually end up at the end of corridors.

Corner aspect

Corner apartments are even better, with views out over two or even three different directions. More opportunity for windows, sunlight, ventilation and privacy; the corner apartment is highly desirable and therefore often quite expensive. Experimentation by Omrani et al. (2016) shows that for an apartment building in the hot environment of Brisbane, Australia, the simple act of having cross-ventilation can provide a comfortable thermal setting up to 70% of the time by natural ventilation alone, while a single-aspect apartment with windows on only one side may achieve that thermal comfort only 1% of the time on a hot summer's day. Corner aspect apartments will obviously have considerably more external envelope than a single-aspect apartment, and as external envelopes are expensive, this means that they will inevitably be more expensive to build.

Penthouse

Rather than using the roofs for air-conditioning plant and services, as is often done on commercial office buildings, penthouses provide luxury living at the top. The penthouse apartment at the top of the apartment building has no neighbours on either side and no neighbours above – meaning it is more like living in a detached house in the suburbs, except for the view – unrestricted for 360 degrees. Roofscapes are often varied, with external landscaping of decks and soaring roofs.

You could argue that the new breed of building like Manhattan's 'super-skinny' apartment blocks with just one apartment per floor are all penthouses, with a single central core serving every apartment. In truth, these are a financial aberration caused by obscene amounts of wealth and are not a solution to the world's housing problem. This book is going to ignore them. They're part of the problem, not part of the solution.

Skip-stop apartments

There's only really one way to have large numbers of dual-aspect apartments, and that's the method that Corbusier found with the Unite d'Habitation in Marseilles (Le Corbusier, 1952): two-storey (duplex) apartments with internal corridors every other floor. It's a lot more complicated to design and build but creates apartments that are more flexible and responsive to the surrounding urban environment. The internal corridor to every second floor was deliberately wide so that it could function as an internal 'street', for children to play in the common space, without the risk of cars.

While it means that there are only corridors and lift access every second or third floor, it also means that every apartment requires an internal stair. Admittedly this is inefficient in terms of space planning and restricts access to the whole apartment to able-bodied people only, but it does create an exciting mix of apartments with varied forms of layout, making for a very interesting building to live in. Other examples of skip-stop apartments include the brilliant but imperilled Gordon Wilson flats in Wellington (arch: MoW: Gordon Wilson/Ernst Plishke, 1959), Peabody Terrace in Harvard, Massachusetts (arch: Jose Sert, 1964), the Trellick Tower in London (arch: Ernő Goldfinger, 1972) and also the afore-mentioned Pruitt-Igoe project in St Louis, Missouri (arch: Minoru Yamasaki, 1954).

Single-core tower blocks

Immediately after the destruction unleashed by the Second World War, there was an urgent need in Europe for new replacement housing estates. The tower block was a popular solution with many Council housing departments bringing Corbusier's 'Radiant City' vision to life. Depending on how you look at it, tower blocks are either a great success or a tragic crime. Your view may differ, if you have actually had the opportunity to live in one.

In working class areas like the heavily bombed and war-ravaged East End of London, towers were built extensively in the 1950s and 1960s. In theory this should have been a wonderful answer to the problem of rehousing the population, but in reality it was not the glorious vision its planners had imagined. The tower blocks of Britain were often inadequately managed, became run down and were much hated; most have subsequently been demolished. Since the Grenfell Tower disaster in west London, where 70 people died in an inferno caused by bad cladding and poor fire risk management, tower blocks are now even less popular in Britain (Figure 4.2).

Most tower block housing takes essentially the same approach of a single vertical access core with four or more single-level apartments per floor. This manifests as a square tower often about 20 floors tall, usually with just a single staircase and a minimum of two (hopefully working) lifts. The public space on

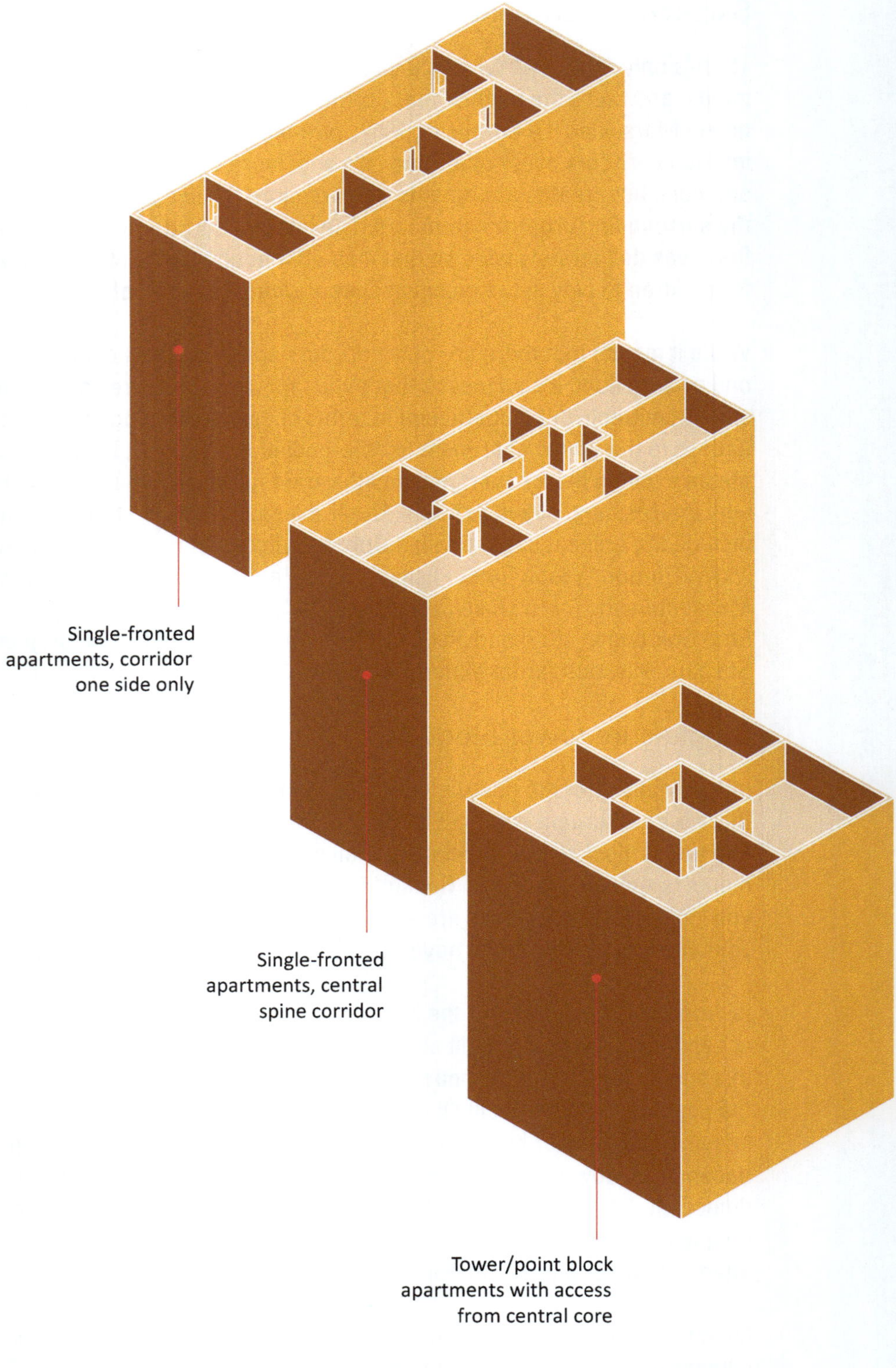

Figure 4.2 Tower block circulation: single-loaded corridors versus double-loaded.
Source: Hayes/Marriage.

each floor is limited to a small lift lobby with apartments opening directly off that: no lengthy corridors. The efficient manner of the tower planning means that while the internal circulation is tight and concise, most of the apartments have good external views from behind the extensive external cladding. If they had been better maintained, perhaps more of them would still exist.

The tower block housing form has been taken on as the key housing solution in crowded cities such as Hong Kong and Singapore. Indeed, for most of the population in Hong Kong, tower blocks are the only way to live. The plan is conceptually similar: a single central core with twin stairs and a number of lifts, with extensive external envelope walls. However, their planning is rather complex. Under the watchful eye of the Hong Kong Housing Authority (HKHA), tower blocks are usually 40 storeys high, constructed in a cruciform shape with 16–20 flats per floor. The HKHA know that management and maintenance of the blocks is crucial and so they are, on the whole, rigorously well-maintained. Hundreds of these cruciform concrete towers are packed tightly onto Hong Kong Island, Kowloon and the New Territories in an extremely dense urban environment. The flats are unfortunately uniformly tiny: just 15m^2 (161 sq ft) each for a 'New Harmony' unit aimed at a married couple. A single tower block in the New Territories of Hong Kong can therefore accommodate 1,600 or more people, with new estates each containing hundreds of similar towers. Density is high and the resident population is crammed in.

But neither the towers of Hong Kong nor the ruined UK Council estates of the 1960s and 1970s should provide our model for the future. Modern apartment towers are one of the answers to the world's current and future housing problems and so the new buildings need to be designed well and constructed to rigorously high standards. Good housing is quality, long-lasting housing, no matter whether it is in the sky or on the ground.

Slab blocks

For many apartment sites, the best option is a rectangular block, prioritising the majority of views out one side or the other: mainly single-aspect apartments off a single-loaded or double-loaded corridor. Another left-over planning concept from Corbusier, this is a form of housing which can be monotonous and boring, or depending on treatment, can also be innovative and exciting. The key to transforming a flat façade into a more responsive and exciting façade will be discussed further in Chapters 11 and 13. It will take hard work and careful attention but will pay off admirably.

Courtyard housing

Apartments arrayed around a central courtyard is a traditional form of housing for many parts of Europe, originally used for sheltering the horses or corralling the carriages. Like the *caravanseri* of the Middle East, where the courtyard created a sheltered garden paradise with greenery, fountains and occasionally camels and donkeys, the courtyard apartment developments of today should strive to keep this central area reserved for the building inhabitants to enjoy as semi-public space. It's important to dedicate this central communal area to people and not to the camel's and horse's successor: the automobile.

Common in cities such as Berlin, courtyard housing presents a solid wall of living rooms to the front of the apartment block, while a large entryway through the front façade to the courtyard permits the inner circle of apartments to have windows and public spaces facing inwards. An enclosed courtyard will provide not just air and light but will also provide a high degree of security and an excellent place to site the circulation. Use of a central courtyard allows a layer of personal privacy within large ring-fenced blocks of urban land. Crucially, it also enables a large reduction in noise compared to the streets outside.

In New York, BIG have designed VIA 57 West, a large apartment building around a large central courtyard on the water's edge of Manhattan Island. With the top of the building carefully cut away at a steep slant, each apartment enjoys a stunning view (architect: BIG, 2016). Despite the radical and beautiful external form of VIA 57 West, the interior planning is still fairly standard. Several different cores rise vertically, with internal corridors leading off to single-aspect apartments either side. Some apartments look out onto the street, while the other half face into the courtyard. The internal courtyard is carefully landscaped to offer zones for relaxation and exercise activities: human recreation, not vehicle storage. Crucially, the ground floor is devoted to retail purposes so that the scheme becomes an integral part of the New York urban landscape.

Urban design issues of apartments

Buildings should not be considered as individual elements in the city but as part of the urban fabric. Buildings and their surroundings affect each other both in performance and energy exchange. Existing housing problems result from a complex interrelation of social and economic factors. Our lives have been changed rapidly because of advancing technologies. They have changed the way we work, live and socialise with each other (Newman, 1961). The building industry has lagged behind these changes, mostly because changing direction in a built environment is hugely time consuming and expensive. While in the past there has been a tendency in low-rise, sprawling cities to simply renovate existing houses, in the future there will be an imperative to densify by demolishing older single storey housing and replacing with apartment buildings.

Apartment life is very important because residents share space vertically and horizontally. Rapid urbanisation shifts citizen's lifestyles from living in simple detached and stand-alone houses to medium-rise or high-rise buildings. Apartments are usually defined as cost-effective smaller spaces which share facilities with others, in contrast to the more private and larger spaces which are offered by (usually) bigger and more expensive stand-alone houses.

Some key issue of high-rise buildings are summarised below:

- Finding the optimal urban site: it is very important for city planners and urban designers to allocate the proper site for high-rise buildings. However, finding the optimal site seems unachievable. Altitude, wind temperature, wind orientation, slope, seashore breeze, flood and finally solar orientation of the site are the main important items for choosing the environmentally best location of the site (Golany, 1996).
- Residential social and psychological aspects: noise and security issues can make towers unpleasant to live in. Frequent contact with neighbours can aggravate problems even more. High-rise buildings are not a favourite place for some parents with children and the perceived fear of crime is very high among tall building dwellers (Gifford, 2007). However, curiously, neighbourhood satisfaction in high-rise buildings is much better than in low-rise neighbourhoods (Kearns et al., 2012).
- Traffic problems: although high-rise buildings shorten the distance between destinations, at peak times, this can increase the traffic jams and overload streets and public transportation. Walking, however, can thrive.
- Shadow creation: high-rise buildings block sunlight in the cities, and therefore, the location and distance from other buildings are quite crucial.
- Visual effects: tall buildings stand out as a feature in the surrounding environment and when clustered together, they can act as nodes in the cities. They become responsible for the context of the city visually, socially, functionally and environmentally.
- Fire evacuation problem: fire evacuation not only needs to be considered in the building but also in the streets around the building. Access for fire engines needs to be considered along with access to water and the overall height of the buildings.
- Micro-environment design: due to their inherent form, high-rise buildings change the microclimate of surrounding areas. The building's layout can change the wind regime and energy absorption of the area. These effects need to be considered in urban planning.
- Façade design: with façade design, architects need to provide a good view, natural lighting and good orientation for each apartment (Niu, 2004). Since each floor consists of different dwellings, achieving all these goals together is difficult and needs skilled attention.
- Providing a balcony for each apartment: having a balcony is necessary for resident wellbeing but can create visual issues from mismanagement and can impose more loads on the building structure.

- Providing natural ventilation: it is very necessary to freshen the inside air; however, this is highly dependent on the rate of pollution of the neighbourhood microclimate.

In the next chapter, we explore the inside of the apartment.

Works cited

Ballard, J.G. 1975. *High-Rise*. London: Penguin.

Bristol, Katherine. May 1991. "The Pruitt-Igoe myth." *Journal of Architectural Education, 44*(3), 163–171. https://doi.org/10.1080/10464883.1991.11102687

Cohen, Jean-Louis and Benton, Tim. 2008. *Le Corbusier le grand*. London: Phaidon.

Gifford, R. 2007. "The consequences of living in high-rise buildings." *Architectural Science Review, 50*(1), 2–17.

Golany, G.S. 1996. "Urban design morphology and thermal performance." *Atmospheric Environment, 30*(3), 455–465.

Heathcote, Christopher. May 2020. "When modernist Britain fell down." In *Quadrant*, 82. Sydney: Quadrant.org.au.

Kearns, A., Whitley, E., Mason, P., and Bond, L. 2012. "Living the high life'? Residential, social and psychosocial outcomes for high-rise occupants in a deprived context." *Housing Studies, 27*(1), 97–126.

Newman, Oscar. 1961. *New Frontiers in Architecture*. A summary of the 1959 International Congress of Modern Architects conference in Otterlo. An early exploration of the effects of architectural design on perception and behavior. New York: Universe Books.

Newman, Oscar. 1972. *Defensible Space*. New York: Macmillan.

Newman, Oscar. 1996. *Creating Defensible Space*. New York: U.S. Department of Housing and Urban Development Office of Policy Development and Research. Available at: https://www.huduser.gov/Publications/pdf/def.pdf

Niu, J. 2004. "Some significant environmental issues in high-rise residential building design in urban areas." *Energy and Buildings, 36*(12), 1259–1263.

Omrani, Sara, Garcia-Hansen, Veronica, Drogemuller, Robin, and Capra, Bianca. 2016. "Thermal comfort evaluation of natural ventilation mode: case study of a high-rise residential building." *In Fifty Years Later: Revisiting the Role of Architectural Science in Design and Practice: 50th International Conference of the Architectural Science Association 2016*, edited by J. Zuo, L. Daniel, and V. Soebarto, 1–9. Adelaide, SA: The Architectural Science Association and the University of Adelaide.

05

Chapter Five

Sunshine, daylight, healthy living and sustainability

Sunshine, daylight, healthy living and sustainability

Guy Marriage

DOI: 10.4324/9781003123873-5

Morning, sunshine!

I am lucky enough to live in part of the world with a moderate climate: never too hot or too cold, although it is a trifle windy at times. Around the globe, many other cities will have more extreme temperatures: boiling hot or freezing cold, and in some cities, both of these extremes in the same year. This chapter, therefore, needs to be read bearing in mind whether it is appropriate for your climate or not. For most apartment buildings around the world, follow these guidelines:

Basic rules

- Always face living spaces towards the sun – allowing free solar heating energy right into the apartments.
- In super-hot climates, ensure that there are deep reveals or shading to windows to allow shelter from the sun.
- Create external private space with a usable balcony for every apartment.
- Ensure excellent thermal insulation to all external walls.
- Design in excellent acoustic separation to all inter-tenancy walls.
- Gain as much usable space as is possible – don't waste space on hallways and corridors.

Orientation

A truly urban building site will be within the city, probably on land that has been developed before, within a matrix of existing neighbours. Typically, an urban site will have other buildings on three of the surrounding four sides, with the only 'free' side being the street: if you are lucky your site is on a street corner, with two free sides exposed to the street.

There are some places in the world where the key consideration may be to keep away from the sunshine, e.g. Dubai, Cairo and other hot places near the equator. For most other places, a moderate amount of sunshine is important, not just for the inhabitants to soak up some vitamin D but also to drive away cold from dark damp corners.

Solar orientation is always a key consideration. The sun always rises in the East and sets in the West wherever you are in the world, but where the sun is at midday varies considerably. In the Northern hemisphere the sun at midday comes from the South, while in the Southern hemisphere the sun at midday comes from the North. In the tropics and near the Equator, the sun at midday comes from directly above, while nearer the Poles at the extreme ends of the Earth, the sun is nearly always arriving horizontally, if at all. An apartment building in Alaska is going to be a very different building from one in Ecuador: one city plan does not fit all.

The ideal orientation for an apartment building in a moderate or cold climate is for all apartments to face the sun – particularly, try to ensure any habitable rooms (living rooms, bedrooms) benefit from solar gain – including at dawn and dusk. Presenting only one façade towards the sun means that the other side of the building will be in the shade, as well as ensuring that the property behind will have their sunshine heavily blocked out for most of their day. How nice do you want to be to the neighbours? Well, perhaps as nice as you would want a neighbouring property to be to you? We're all in this together! If apartment blocks are oriented North–South, this allows one side of the building to have sunshine in the morning and the other half to get afternoon sun, while also allowing the sun to access the neighbouring site behind.

Morning sunshine (i.e. rooms facing East) is good for waking up to, or to have streaming into the kitchen early in the day. Avoid having the kitchen facing West – it's hot enough at the stove without the sun cooking you as well. Late afternoon sunshine (i.e. rooms facing West) is much more enjoyable in the living room or out on the balcony. Sitting outside celebrating the sunset with a drink in hand after work, as you look out over the city is much more enjoyable while being basked in the sun's rays, rather than sitting in the shade.

With people living in a city, midday sun is arguably less important throughout the working week. Traditionally, most apartment buildings will have homeowners working in offices during the day and coming home at night. Of course, this habit of physically going to work may change in a post-Covid world. We should expect a greater level of working from home in the future, or at least a need for defined home offices so that the children cannot wander into while you're in a video-conference (BBC News, 2017).

Urban sites

Each city tends to have its own size, shape and orientation of its city blocks; an excellent and comprehensive resource on city blocks is the book *Urban Grids* (Busquets et al., 2019). As an example, Manhattan Island in New York has city blocks set out in a regular pattern with over 200 streets crossing East–West between the nine main avenues running North–South. Typical block sizes in New York are 244m wide × 61m (famously 800′ × 200′), broken down further to 7.6m × 30.5m (25′ × 100′) standard sized sites of 232m^2 (2,500 sq ft). The small cross-street frontages typically face into or away from the sunshine (i.e. North or South) and are enclosed on each long East or West side by their neighbours. Most New York apartment buildings are on sites that have been amalgamated from a large number of smaller sites, but very few cover a whole city block. BIG's New York housing project VIA 57 West takes up an entire city block, one of the few new schemes to do so on.

By comparison, Chicago has almost square city blocks of 100m × 120m (328′ × 394′), while Los Angeles has suburban and city blocks based around a 100m × 180m (328′ × 591′) grid, but oriented at 90° to those in New York.

Older cities in Europe are far more varied, with irregular roading patterns dating from ancient Roman times in many cities, including London. Many European cities also have a new grid layout for the more modern parts of their city – most famously Barcelona's 'new city' blocks are an octagonal doughnut of space measuring 113m × 113m (371′ × 371′). City sites are more challenging to design but are ultimately more satisfying to complete. Suburbs of cities may situate apartment buildings in open green field land, allowing tower blocks to take almost any form but will then require occupants to have a car in order to access the jobs and civic life in the centre.

An apartment building I helped design, and still live in 20 years later, is an old warehouse: full of industrial character. It has an unusual orientation – the 60m (196′) wide street frontage faces the sun but is only 16m (53′) deep, a total of 960m^2 (10,333 sq ft) per floor. This permits every one of the eight apartments per floor to face into the sun, all day long. The key to the continued success of this building is that there are a variety of different-sized apartments ensuring a good mix of different residents: singletons, double incomes (dinkies), families with children, retirees, students, including nurses, council workers, film-makers, musicians and at least one architect. The variable size of the apartments has also allowed companies to be started up in living rooms, all contributing to the activity in the building – a graphic design office, an architectural studio (ours) and much to our surprise, for a while, a high-class brothel (not ours). We clearly have good sound-proofing – just as well, as I live next door to a trombonist in the NZSO! Nearly every apartment here has a decent-sized 5m^2 balcony – and all the top floor apartments each have a 12–15m^2 roof garden as well. On a sunny weekend the façade comes alive with residents animating the building in their leisure time; the rest of the time, the action is on the inside. A good apartment building is a joy to live in, but even more so if you cannot hear your neighbours.

Ventilation

There is one distinct drawback to single-aspect apartments however: the ability for the apartment to rely on natural ventilation. If windows are only on one side of the apartment, it is difficult to get natural breezes to flow through the rooms within. One way to encourage a natural flow of air is to open the front door and let the air-pressure find an exit out the rear of the building. Clearly this is an undesirable solution, as the front door to the apartment is part of the fire compartmentation and so wedging the front door open would be unsafe for fire as well as security reason (Figure 5.1).

A dual-aspect or corner apartment works so much better at enabling air to move naturally through (Omrani et al., 2016). Natural ventilation is highly desirable, reducing electricity costs (no need for air-conditioning) with the cooling effect of moving air currents being a great way to stay at a comfortable temperature. Not all cities have climates where this is possible: New York, Chicago, Moscow, and Beijing all have summers where it can be blisteringly hot as well as winters where the snow piles up, meaning that windows are unlikely to be left open all year round.

There are some disadvantages of having opening windows and relying on natural ventilation, of course, especially if you live near a major road. Years ago, when I lived in the centre of London, the diesel dust from taxi exhausts used to find its way into the living room even with the windows closed; plenty of urban dwellers still have that problem. But it is not just pollution that is the problem – one of the biggest issues for a city dweller is noise, from both inside and outside the building, by day and by night. There's more on this subject in Chapter 12.

Heating

While 150 years ago apartment buildings were heated by coal-burning fires, these days you will need another source of heating and cooling. In North America, it is common for all apartments to be fully air-conditioned via heating vents powered by an oil-fired boiler down in the basement or up on the roof, although now new buildings now are more likely to use gas-fired or wood-pellet burners as a heat source. The central ventilation system reticulates the warm air throughout the building and into apartments via a system of carefully fire-rated ducts and dampers. In Britain it is more common to have a continuous supply of hot water from a central boiler to a series of cast-iron radiators, each sited by a window. By contrast, in Australia and New Zealand, there is no central heating, it is more common for each apartment to have its own small heat-pump air-conditioner, fixed to the external face of the building or often on an external balcony. None of these are particularly good solutions for energy efficiency.

We need to be a lot more clever about heating and cooling our buildings and must strive to eliminate using fossil-fuels to heat and cool them. Coal, oil and gas are out. Wood-pellets can provide heating from a renewable source, but solar energy, wind energy and geothermal ground heating are effectively free. These are surely the heating sources of the future. Insulate well, orient towards the sun, and use the hive effects of neighbours to stay warm. Apartments, if designed well, should always be so much more

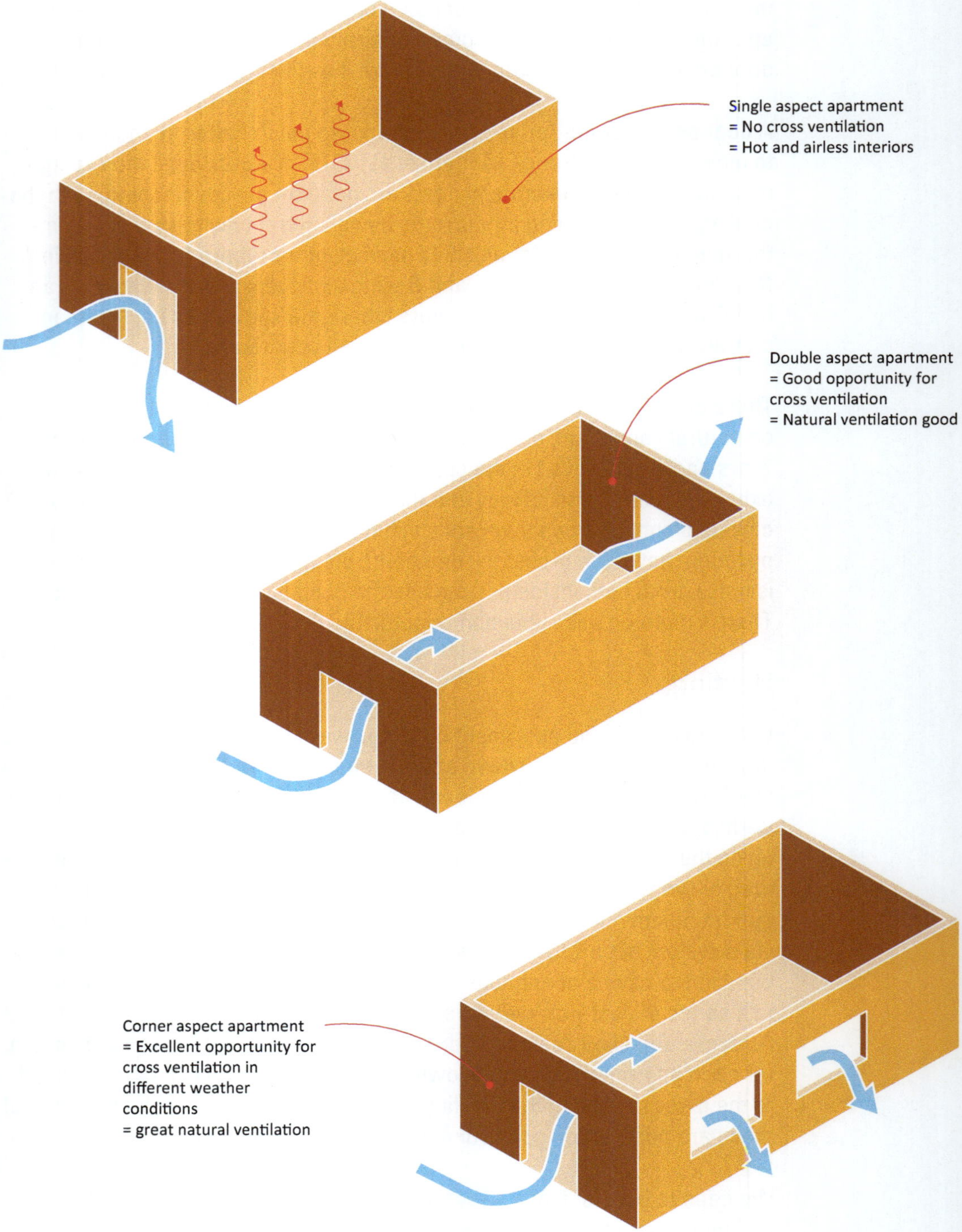

Figure 5.1 Fresh air and ventilation.
Source: Hayes/Marriage.

warm and comfortable than single houses. As a box-like structure, a house typically has four sides, and thus a total of six surfaces to insulate. By comparison, an apartment may often just have one or two exposed sides – far easier to keep warm.

Solar power

Suburban houses can easily incorporate solar photo-voltaic (PV) panels onto their roofs because (generally), the roof area is large enough to provide all their electrical needs. Apartment buildings are not nearly so easy to incorporate PV panels, being generally tall and thin rather than low, flat and wide. The limited roof space on a tall building means it is unlikely to be able to power the entire building from its roof panels. So tall elevations facing the sun make more suitable surfaces for fixing PV panels, and the panels can be designed into the facades, although neighbouring buildings may cause unwanted shading.

By using places such as balcony fronts, *brise soleil* and solar shading awnings, there is the chance for a double success: providing not only solar power but also solar shading at the same time. Wind speeds high up on a tall building can be extreme and can easily tear away any flimsy structure, so sturdy construction is needed, and maintenance access must be designed in with wiring connection points easily reachable from the apartment. Due to the requirements of the wiring rigs and voltage transformers needed, the best result for a PV solution is to integrate solar generation and storage batteries into the design early on, not to try and retrofit a solution later. As solar PV panels continue to improve in efficiency, expect them to become an integral part of the façades and roof. New generations of solar PV can be integrated into a thin film on the glazing, a potentially great solution when costs come down.

Wind power

There has been limited uptake of wind energy generators permanently affixed to buildings so far this century, with some examples being more notable by their gimmicky nature than by true successes as power sources. The Strata office tower at Elephant and Castle in London prominently features three wind turbines set into the peak of the façade as a strong symbol for the building, although actual power output is reasonably small (only 8%) compared to what is required for the whole building (Glancey, 2010). Admittedly, London is not an excessively windy city (usually), but perhaps windier cities can achieve better results. A well-known city with a reputation for being windy is Chicago, but it actually achieves only an average wind speed of just 16.6 kph (10.3 mph). Chicago is outranked by several other cities, including Dodge City in Kansas, which averages speeds of 22.3 kph (13.9 mph) Table 5.1.

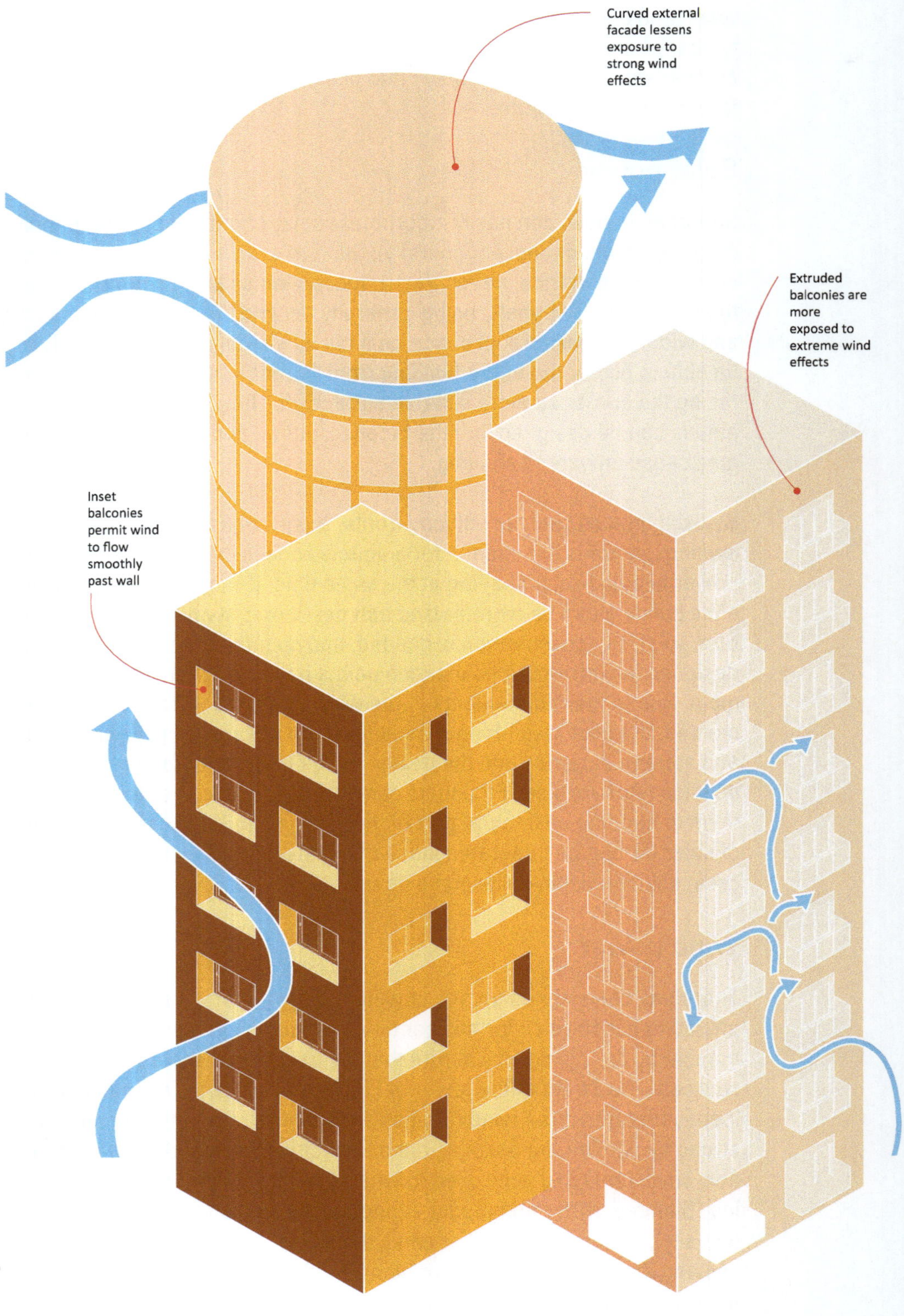

Figure 5.2 Wind effects of balconies on buildings.
Source: Hayes/Marriage.

World's windiest city

Wellington (New Zealand) frequently tops the list of windy cities with recorded gusts over 100 kph (60 mph), although average wind speeds are a mere 27 kph (17 mph). The wind speed record there was set in 1962 where a wind gust peaked at 247 kph (154 mph) on a nearby hill (Mathieson, 2015).

So, are buildings in Wellington covered in wind turbines? Actually, no, but there are a lot of massive wind turbines on hillsides a small distance outside the city that supply enough electricity to power the entire population. Famously, the wind almost never stops blowing – it is a very efficient wind farm. All knowledge of local wind conditions is important not just for potential energy generation, but also for the design tolerances of the façade system to withstand the SLS and ULS (Figure 5.2).

A key factor of any turbine installation will need to be the avoidance of any vibration transmission, primarily during night-time wind generation when residents are trying to sleep and their senses are more sensitive. While tall three-bladed wind turbines are very powerful, they can also be noisy up-close and act as a giant vibration source with a low buzzing hum, or (to some) a deeply irritating throb. Not something that you want in a city building then, especially with giant blades thrashing round and the wing tips reaching speeds of 290 kph (180 mph). Urban residential turbines are generally small, either vertically aligned or horizontally mounted (Cace et al., 2007), with experimentation for options like the Liam F1 Urban Wind Turbine (Coxworth, 2014). Most recently, at the time of publication, the prospect of waggling, bladeless, vertical wind turbines dubbed 'skybrators' shows that the possibilities of wind energy generation continue to be explored (Ambrose, 2021).

Geothermal ground energy

The thermal mass of the earth beneath our feet provides a stable and constant energy source. While above-ground temperatures can fluctuate considerably, dependent on the climate and environmental conditions, far below the surface the temperature tends to stay fairly constant. In other areas, a steady supply of cool water may be used to provide a constant source of coolant. In most other areas of the globe however, the constant temperature of the earth can be utilised via a heat exchanger to gain a source of heating/cooling that can be used for buildings. Pipework arrays are dug into the ground and filled with coolant, which is circulated back into the building and the coolant run through a heat exchanger. In some areas such as Reykjavík (Iceland) and Rotorua (NZ), naturally heated geothermal hot-water springs can provide a reliable source of heat, often captured by local housing projects. More expensive to install but way cheaper to run, geothermal systems can also be used to cool buildings: BIG's VIA 57

Table 5.1 Wind Speed conversion table

Kilometres/ hour (kph)	Miles/ hour(mph)	Meters/ second (m/s)	Feet/ second (ft/s)	Knots (kn)	Beaufort scale	Beaufort description	Conditions on land
5	3.1	1.38	4.55	2.7	1	Light air	Smoke drifts
10	6.2	2.77	9.11	5.4	2	Light breeze	Wind felt on face, leaves rustle
15	9.3	4.16	13.67	8.1	3	Gentle breeze	Leaves and small twigs in constant motion
20	12.4	5.55	18.22	10.8	4	Moderate breeze	Small branches move, dust raises
30	18.6	8.33	27.34	16.2	5	Fresh breeze	Small trees begin to sway
40	24.8	11.11	36.45	21.6	6	Strong breeze	Large branches move, umbrellas used with difficulty
50	31.0	13.88	45.56	27.0	6	Strong breeze	
60	37.3	16.66	54.68	32.4	7	High wind, near gale	Whole trees in motion, inconvenience when walking against the wind
70	43.5	19.44	63.79	37.8	8	Gale/fresh gale	Wind impedes progress
80	49.7	22.22	72.90	43.2	9	Strong/severe gale	Slight structural damage to roofs
90	55.9	25.0	82.02	48.6	10	Storm/whole gale	Trees uprooted, considerable damage: seldom experienced inland
100	62.1	27.77	91.13	54.0	10–11	Violent storm	Very rarely experienced: widespread damage
150	93.2	41.66	136.70	81.0	12	Hurricane-force winds – devastation	Devastation on land; sea completely white with foam and spray
200	124.2	55.55	182.27	107.9	*	*	
250	155.3	69.44	227.83	135.0	*	*	

Note: * = off the scale.

West project cools the building via water collected from under the Hudson river running alongside Manhattan Island.

Community heat and power

Instead of heating or cooling building projects individually, sometimes several different buildings seek to share an energy source together. Sometimes this may be utilising the heating given off by a nearby industrial heat source. For instance, an electricity generating power station requires cooling and so takes in cold water and feeds out hot water (one of the few perks of living near a nuclear power station). If that is not available, then a Community Heat & Power project (CHP) will allow one small neighbourhood energy generation plant to create a source of warmth – perhaps by burning biomass or community rubbish. BIG's new ski-slope/rubbish transfer station/power generation plant in Copenhagen is one innovative example. Here at Amager Bakke (aka CopenHill incinerator) the rubbish is collected, waste is burnt to generate heat and electricity, and the resulting slope on the building can also be used as a ski slope (BIG, 2017). However, nothing, no matter how innovative, goes ahead without at least some issues (Madsen, 2019).

Works cited

Ambrose, Jillian. 2021. "Good vibrations: bladeless turbines could bring wind power to your home." *The Guardian*, 16 March 2021. Available at: https://www.theguardian.com/environment/2021/mar/16/good-vibrations-bladeless-turbines-could-bring-wind-power-to-your-home

BBC News. 2017. "Children interrupt BBC News interview – BBC News." 10 March 2017. Available at: https://www.youtube.com/watch?v=Mh4f9AYRCZY

BIG. 2017. ARC CopenHill/Amager Bakke. Available at: https://big.dk/#projects-arc

Busquets, J., Yang, D., and Keller, Michael. 2019. *Urban Grids: Handbook for Regular City Design*. Harvard: Oro Editions.

Cace, J., ter Horst, E., Syngellakis, K., Niel, M., Clement, P., Heppener, R., and Peirano, E. 2007. "Urban Wind Turbines – Guidelines for small wind turbines in the urban environment." *Intelligent Energy Europe*. Available at: http://www.urban-wind.net/pdf/SMALL_WIND_TURBINES_GUIDE_final.pdf

Coxworth, B. 2014. "Screwy-looking wind turbine makes little noise and a big claim." *New Atlas*. 27 May 2014. Available at: https://newatlas.com/the-archimedes-liam-f1-urban-wind-turbine/32263/

Glancey, Jonathan. 2010. "Spin city: strata tower." *The Guardian*, 18 July 2010. Available at: https://www.theguardian.com/artanddesign/2010/jul/18/strata-tower-london-green-architecture

Madsen, John. 2019. "A Danish fiasco: the Copenhagen incineration plant." *Zero Waste Europe*. Available at: https://zerowasteeurope.eu/2019/11/copenhagen-incineration-plant/

Mathieson, K. 2015. "Where is the world's windiest city? Spoiler alert: it's not Chicago." *The Guardian*, 15 October 2015. Available at: https://www.theguardian.com/cities/2015/oct/15/where-world-windiest-city-spoiler-alert-chicago-wellington

Omrani, Sara, Garcia-Hansen, Veronica, Drogemuller, Robin, and Capra, Bianca. 2016. "Thermal comfort evaluation of natural ventilation mode: case study of a high-rise residential building." In *Fifty Years Later: Revisiting the Role of Architectural Science in Design and Practice: 50th International Conference of the Architectural Science Association 2016*, edited by Zuo, J., Daniel, L., and Soebarto, V., 1–9. Adelaide, SA: The Architectural Science Association and the University of Adelaide.

06

Chapter Six

Structure and materiality

Structure and materiality

Nabil Jose Allaf

DOI: 10.4324/9781003123873-6

Introduction

High-rise residential and commercial buildings share many similarities, in particular, the structural systems ensuring gravity and lateral loads resistance. From a purely technical approach, any structural system presented in *Tall: The Design & Construction of High-Rise Architecture* (Marriage, 2019) is appropriate. However, as these two types of buildings respond to distinctive users, their purposes and functions differ. Their differences influence the structural layout and materiality being used, leading to some structural systems being more suitable than others. Some of the main requirements that designers should consider for apartment buildings are:

- establishment of definitive boundaries
- acoustic privacy
- fire-proofing

In contrast to high-rise commercial and office buildings, where the structural system is aimed at providing maximum freedom and flexibility, apartments need a clear and definitive spatial layout delimited by both structural and non-structural boundaries. To respond to this demand, the use of structural systems such as shear walls, rigid frames and bundled tube appear relevant. The choice of one system over the other, or in combination, depends on the desired height for the building to reach as well as the respective architectural qualities each structural system provides.

Structural components

Rigid frame

Rigid frame is one of the primary structural systems to consider when planning a residential high-rise building. It consists of a combination of columns and beams connected through rigid connections which allows the bending moment to be transferred from the horizontal members to the vertical ones. In earthquake-prone areas, the rigid connections would also need to display sufficient ductility in order to face lateral drifts without snapping. This system is capable of providing sufficient structural stiffness to reach up to 30 storeys; beyond that height the sway would be considered excessive. The square or rectangular form of the system simplifies the internal layout and provides great flexibility in terms of planning. In addition, the structural grid can be adjusted depending on the designers' requirements, as long as structural stability is maintained. This is mainly expressed by the distance between the columns which, as it increases, leads to a greater span of the beam and therefore of its depth. Careful structural layout is therefore required to ensure adequate floor to floor height.

Rigid frame structures can be made out of reinforced concrete, steel or mass timber. Each one requires careful detailing, with each materiality possessing a variety of possible rigid connections. In the case of reinforced concrete, the rigid connection is made by strongly tying the reinforcement bars of the beam to the ones of the column, using a large amount of steel ties– called stirrups. For steel structures, the rigid connection can be either fully welded or bolted. In both cases, in order to transfer the bending moment from the beam to the columns, the flanges of the beam need to be connected to the flanges of the columns using continuity plates.

Finally, the rigid connection of timber columns and beams, made of glulam or Laminated Veneer Lumber (LVL), need to be carefully undertaken due to the risk of wood fibre splitting. Some options presented by Buchanan and Fairweather (1993) for multi-storey buildings include the use of a circular arrangement of dowels joining double beams to a column, and the use of epoxied bars achieving a strong connection. For further discussion on mass timber, refer to the next chapter and to the Case Studies concentrating on timber.

Shear walls

Shear walls can be designed as part of the building's core, used as the external envelope, or installed as inter-tenancy walls. The use of shear walls made of Cross-Laminated Timber (CLT) or plywood is increasing as the construction of timber high-rise buildings grows, but commonly shear walls remain made of reinforced concrete and can either be designed as precast panels or be cast in-situ.

Precast panels

Precast walls are by definition manufactured off-site, then transported, lifted and installed on site. This system is time and cost-effective in comparison to shear walls cast in-situ. As the walls are built in a controlled environment, the quality of the concrete is also at its best. Precast manufacturing and transportation limitations apply, so the panels are typically one to two storeys high and connected at each floor level.

The most critical aspects to consider when using precast walls are their connections with the horizontal and vertical structural elements. In order to ensure a proper transmission of the shear forces from the floor diaphragm to the precast shear walls, the connections between both elements need to be sufficiently strong and stiff. To achieve this, an effective approach is to have starter bars already embedded inside the wall. When connecting the walls to the floor at the edge of a core, these bars would commonly be angled at 90 degrees, allowing connecting the precast panel to the floor slab once the topping is poured. If the precast panel is used as an inter-tenancy wall, the starter bars would remain straight and connect to both the floor and the above precast wall.

As mentioned, the size of precast panels is predetermined. However, in some cases, longer walls might be required when, for instance, trying to achieve sufficient resistance to horizontal loads. In this case, two or more aligned precast panels need to be joined together in order to act as one single shear wall. An effective method to achieve strong connection between the panels is via cast in-situ joints and consists in each panel having anchored looped steel bars at their joining extremities. Vertical rebars from the walls below would also run through the loops, thus ensuring vertical continuity and transfer of shear force. This junction is then entirely filled with concrete, thus ensuring a strong bond.

Cast in-situ

In contrast to precast panels, cast in-situ shear walls are more labour intensive. They require the contractor to set up elaborate formwork, accurately tie the rebars together and work at height on site. The concrete is then brought to site and poured in the formwork while being vibrated to eliminate any air bubbles. The formwork is removed after three to seven days when the concrete has hardened. The risk with this approach is that if the wall displays structural defects, it will need to be destroyed and started again. However, the advantage of this technique is to create shear walls with greater ductility and strength, thus reaching greater heights. One method of in-situ casting is the use of 'jump forming' or 'slip forming' the core, where the formwork is mechanised to enable it to push itself ever upwards, jacking up from the previous pour. In this method, the core walls are constructed significantly before the rest of the building, allowing the concrete to achieve full strength and hardness before attaching other loads.

Some architects also appreciate the cast in-situ technique as it allows achieving fair-faced concrete walls, meaning walls on which surfaces are exposed for aesthetic purposes. Depending on the formwork's internal facing panel, the wall can display different patterns and textures. One of the most common visual characteristics of these walls, largely expressed by the architect Tadao Ando in his projects, is the presence of holes created by the form-tie points holding the formwork together.

Whether precast or cast in-situ, a reinforced concrete shear wall is particularly relevant when it comes to acoustics and noise control. Its mass makes it an efficient sound insulator, whether the source of the sound is located behind the concrete wall or resulting from an impact on it. While a bare shear wall will perform acoustically very well, additional finishes should be considered to further enhance the acoustic performance. This is particularly important when the shear walls are used as inter-tenancy walls.

Bundled tube

The bundled tube system is another option that can be considered for residential high-rise buildings. It shares similarities with the rigid frame system as the structure resisting gravity and lateral loads is based on a combination of columns and beams. The rigid frame system would more commonly be designed to spread over the entire building's floor, and proportionally distribute the loads to every frames. The bundled tube system combines several single framed tubes, clustered together to operate as a single entity and ensuring proper structural integrity. This system provides greater flexibility for both in internal planning and building forms, as each tube establishes its own perimeter boundary, thus avoiding the use of additional interior columns.

From a planning and internal layout perspective, the system starts as a combination of all the tubes at its base and gradually reduces their number as the height increases. The building gets slimmer as it rises without losing its structural effectiveness. As a result, the usable surface area varies from floor to floor and additional openings start appearing. New opportunities are given to the designers which need to adapt their spatial considerations to the available surface areas as well as consider them in regard to additional natural light entrances and outside views. The structure is therefore the primary spatial divider. The columns of each tube establish a first boundary base inside and along the perimeters. Depending on the required size of the apartments, this first division might be sufficient to define their layouts, each one being an independent flat with the highest tube serving as the core. The flexibility of the system can be further exploited with an internal non-structural layout subdividing each tube into smaller units.

Opportunities also arise from the volumes and shapes of the tubes being used. While the box shape of each tube is commonly referred to when considering the bundled tube system, due to its use in the design of the Willis Tower (previously Sears Tower), alternatives can be considered. As the structural integrity of the bundled system comes from clustering the tubes, potentially any tessellated regular shapes can be used. The variety of geometric figures combined with different grouping combinations therefore offer a large range of dynamic volumes based on tessellated triangular, hexagonal, or diamond patterns (ref. Figure 6.1). However, some configurations repeated over several floors should be carefully considered or avoided so as not hinder the building's structural stability, particularly in earthquake-prone countries.

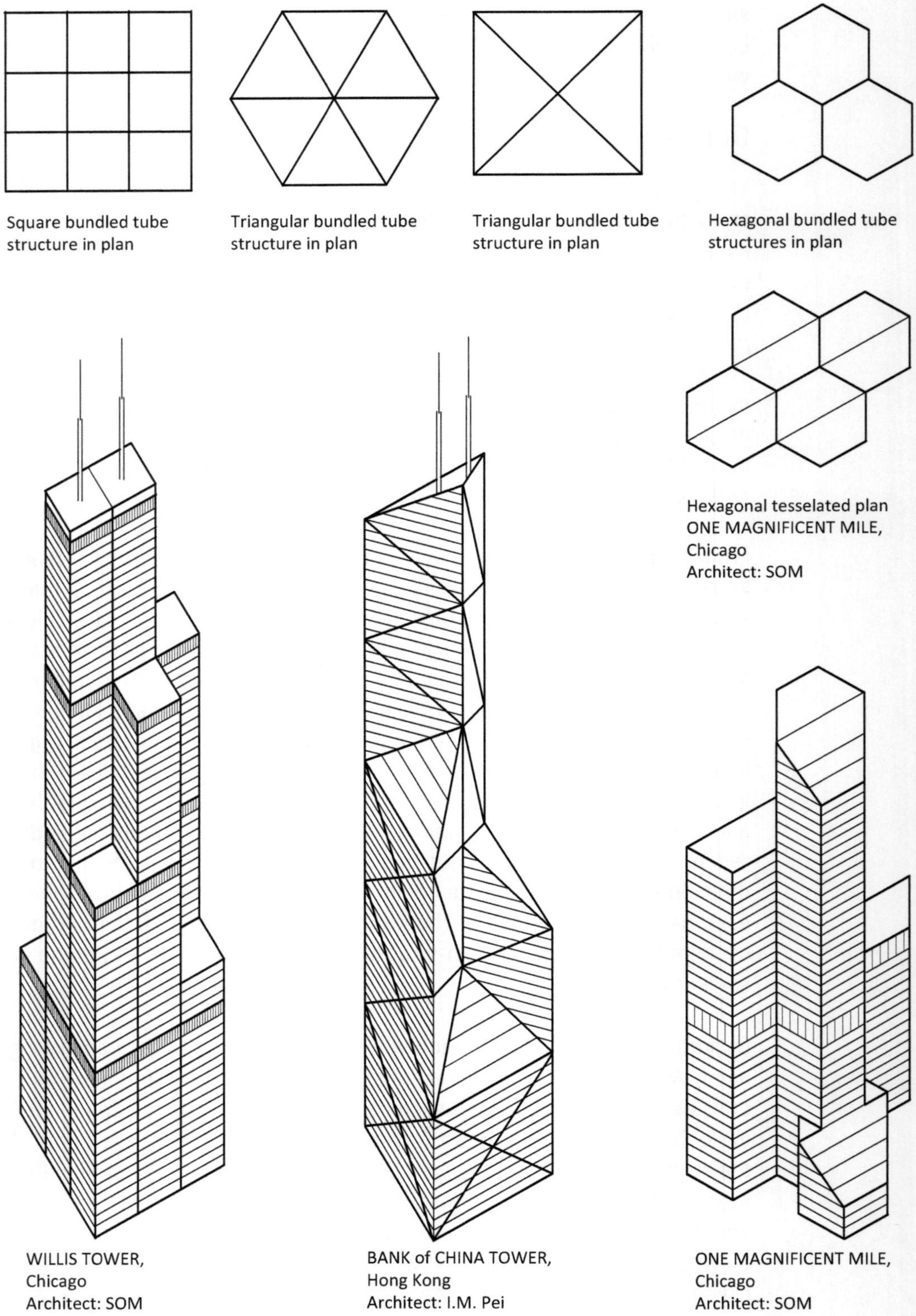

Figure 6.1 Tessellated patterns in structure.
Source: Marriage/Allaf.

The first concern is re-entrant corners, with the building's plan forming a U or L shape. In these scenarios, particularly if the plan creates long lever-arms, large deflections can occur, causing high stress concentration at the corners and possibly damaging them. The other configuration is a diagonal plan, where the tubes only meet at their corners. This approach disregards the effectiveness of clustered tubes. The tubes cannot work effectively as a single entity and the risk is for the junctions to be too weak to resist the stress concentration.

In terms of materiality, steel and concrete remain the primary options when approaching the bundled tube system.

However, the increasing interest for timber is opening up alternatives. One of them is identified as the Urban Timber system (Silva et al., 2013, 2014), which aims to replace the entire building's structure by engineered timber. The framed tubes of the bundled system become a combination of CLT walls, taking over the structural columns, with twin glulam beams connecting them. The beams also support CLT floors which help transfer the loads to the walls. This reinterpretation of the bundled tube system still draws upon the interconnection of the tubes to form a mega structure capable of resisting gravity, lateral loads and overturning moments. The impact of this option on the internal spatial division needs to be carefully considered. The *free plan* design resulting from the traditional bundled tube system using columns is here impacted by the use of walls which have a larger footprint and therefore a greater influence on the spatial delineation and internal layout.

Inter-tenancy walls/demising walls

An inter-tenancy wall is aimed at dividing the area of a floor into distinct apartments. In order to ensure privacy, the wall must perform acoustically while providing sufficient fire resistance and, to some degree, thermal performance. The composition of an inter-tenancy wall varies depending on the materials and structural systems used for the building. While commonly designed as twin planes of brick or blockwork walls in the last century, the preference nowadays is to rely on a framed construction. Inter-tenancy walls can also be part of a building's main structure, as they are with shear walls from precast panels, which separate tenants while also ensuring proper gravity and lateral resistance.

In the case of concrete shear walls, fire protection is provided by the materiality of the selected structural system itself. Steel or timber battens may then be installed on both sides of the wall with insulation located in between them to disguise the concrete within. Linings such as plasterboards are then fixed to the battens. The main issue in using shear walls as inter-tenancy walls comes from the direct contact between all the components being used. As a result, direct impact sound can travel from the lining to the battens through the concrete layer and reach the other side of the wall the same way. In this scenario, additional components should be considered such as light gauge steel resilient channels or rubber mounts, described in a later chapter.

Where shear walls establish a physical and impassable boundary, the use of rigid frames along the separation line of two apartments requires additional components to establish proper compartmentation. In this scenario, the most effective design is the use of twin walls. This approach provides the highest acoustic performance as it features the use of two structurally independent walls installed side by side. Twin walls also easily accommodate services within the partition wall and enable hiding any columns by encapsulating them whilst maintaining a continuous lining.

However, steel cross-bracing within an inter-tenancy wall can cause issues due to its width. Any such bracing needs a plasterboard-clad wall either side, not touching the steel, and so there is an inherent problem with the resulting increased wall thickness. While this may not be obvious to the final residents of the building, any waste of usable floor space is a shame. More commonly, in a concrete-framed building these bracing elements can be achieved to some extent by the use of thin rectangular columns. This is a tactic clearly used in many concrete apartment towers where perimeter columns are stiffened in one direction and then hidden within the walls of the completed apartment. A thin wall of plasterboard on a resilient channel frame then sails cleanly past the column on both sides, permitting only a minimum of space to be lost.

Works cited

Buchanan, A.H. and Fairweather, R.H. 1993. "Seismic design of glulam structures." *Bulletin of the New Zealand Society for Earthquake Engineering, 26*(4), 415–436. http://www.nzsee.org.nz/db/Bulletin/Archive/26(4)0415.pdf

Marriage, Guy. 2019. *Tall: The Design & Construction of High-Rise Architecture*. London: Routledge.

Silva, C.V., Branco, J.M., and Lourenço, P.B. 2013. A *Project Contribution to the Development of Sustainable Multi-Storey Timber Buildings*. Institute for Sustainability and Innovation in Structural Engineering.

Silva, C.V., Branco, J.M., and Lourenço, P.B. 2014. *UT System: A Structural System to Build Taller Urban Timber Houses with Aspired Spatial Flexibility*. Institute for Research and Technological Development in Construction Sciences.

07

Chapter Seven

Sustainable timber future

Guy Marriage

Sustainable timber future

DOI: 10.4324/9781003123873-7

Introduction

This chapter examines the growth of trees and their timber, from the science behind the carbon cycle to the many types of modern engineered timber available today.

Over the last two decades, producing carbon-zero buildings has gone from a conceptual possibility to a reality, becoming a modern standard for world-class environmental action along the way. The importance of understanding carbon has grown, despite financial issues and some people that are wilfully ignoring the science. Let's look at the science first.

Carbon cycle

Our current atmosphere consists mostly as nitrogen (N_2) (78%) with 21% oxygen (O_2) and traces of other gases, mainly (1%) water vapour (H_2O) and a tiny fraction (0.04%) of carbon dioxide (CO_2). It wasn't always like this, 4.5 billion years ago the atmosphere was toxic and unbreathable, as the main gases present were ammonia (NH_3) and carbon dioxide (CO_2). How did it change to what we have now? The answer is: very slowly. A whole lot of leafy greenery was produced over billions of years, growing first in the water as algae and cyanobacteria, and much later on land as plants and trees. Plants drink in water and other nutrients from their roots, absorb CO_2 through their leaves, process the sun's energy through photosynthesis and convert it all into plant growth. The plants then exude O_2 back into the atmosphere in the process of transpiration – an unwanted by-product from this equation. Repeat this process endlessly, and after a very long time, the atmosphere changes into one filled with oxygen.

Animal versus plant

Nature being nature, it finds a use for all the excess oxygen from the equation and so animal life evolved: breathing in oxygen, eating plants and excreting the remains; letting off exhaust gases such as CO_2 and methane (CH_4). Plants feed the animals, while in turn animals feed the plants.

The carbon atoms themselves are the glue in the sandwich, endlessly moving around from one form to another. It is a stable but complex system, a carbon cycle. Life on Earth has been flourishing, with animals, plants and fungi, all inherently part in this cycle. We have been enjoying the relatively steady state we have been in for millions of years, until recently.

Excess carbon is absorbed back into the ocean or rots in the ground, decomposing into an oily black mass of unoxygenated wetland ooze, later covered with layers of sediment on the sea floor. Oil reserves, natural gas, coal and tar-sands are all part of this stored carbon from ancient times. The amount of stored carbon is immense and is a vital part of our planet's health system. It's also a problem for the natural balance if this carbon is released.

Our increased use of oil to create and fuel our modern world means the atmosphere is steadily changing. The amount of CO_2 in the atmosphere has exponentially increased since the 1800s when the Industrial Revolution started, from a relatively steady 250 parts per million (ppm) to over 400 ppm today: 150 ppm is a stratospheric increase over 200 years in a system where changes usually take millions of years. We are currently in the middle of a wave of extinction primarily caused by human over-consumption of resources, with a resulting excess emission of carbon. While the oceans used to be able to act as the moderator of all excess carbon, the process is now increasingly stressed and the higher CO_2 numbers are having an effect on the weather, the oceans, the atmosphere, the animal and plant kingdoms, and, of course, on us.

Carbon calculation equation

The overall calculation of the carbon cycle equation is hugely complex, but it can be reliably stated that both oil-based transport and the building process are key components of this imbalance. The use of both concrete and steel are hugely complicit in the creation of this excess carbon. A building material is needed for mass consumption: one that locks up carbon and does not expel it into the atmosphere (Figures 7.1 and 7.2).

Trees

The key material for the future that does not create this problem is timber: wood, lumber, dead trees. Trees are just tall plants that were grown using H_2O, CO_2, nutrients and sunlight, so when we cut them down to process them, they still contain a mass of carbon. As this is new, fresh carbon taken directly from the atmosphere and not recently from a fossil-based source, we can say that the carbon-based life form of the tree sequesters the carbon away in the timber, locking it up and stopping it re-entering the atmospheric carbon cycle. In today's carbon-conscious world therefore, the use of trees as a carbon sink is growing quickly in importance, both by planting large areas of forest to absorb the over-abundant carbon dioxide problem, as well as the necessary follow-on action of using that timber as a carbon sink, locking the carbon up in timber buildings in a permanent construction. Roughly speaking, one cubic metre of softwood timber will weigh about 500 kg, mostly carbon, taken direct from around 800 kg of CO_2 in the atmosphere.

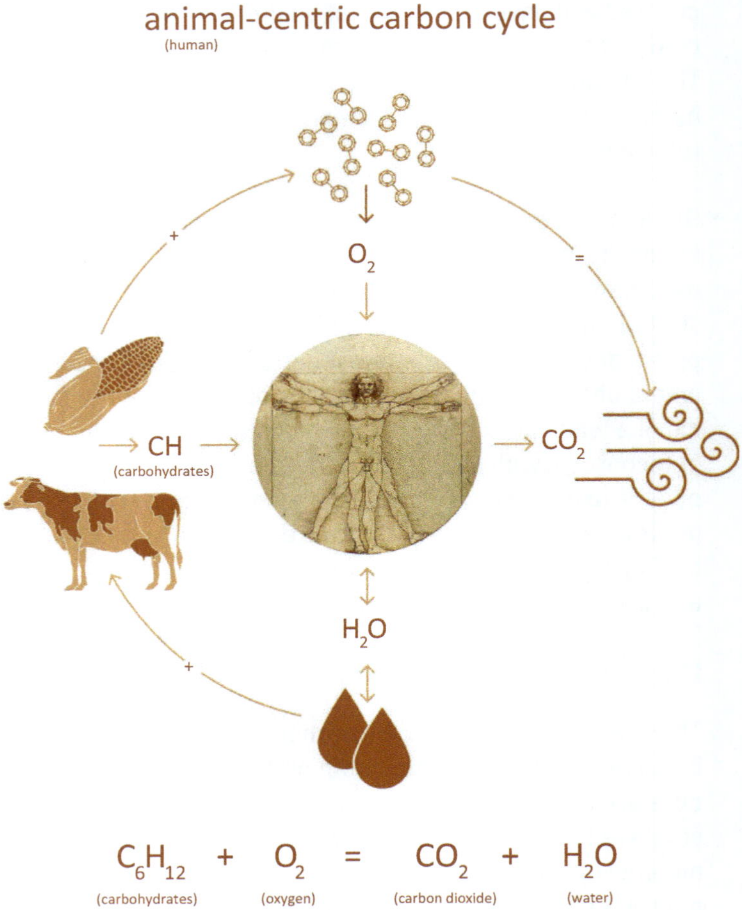

Figure 7.1 Animal-centric carbon cycle.
Source: Hayes/Marriage.

Tree structure

Technically speaking, timber is a series of cells composed of polymeric cellulose strands imbedded in a lignin matrix. Trees grow through a process of the cambium layer creating new layers of cells each year, with more cells created in summer and fewer cells created over winter, hence the visible annular rings of growth. The sapwood beneath the surface transports the sap (water and nutrients) up the trees from the ground to the leaves, while the leaves undertake the photosynthesis that creates more timber. The steady growth of new sapwood allows the heart wood at the centre of each log to form: dense, stable, dead, but strong and highly desirable to build from.

Hardwood trees are mainly deciduous; trees that drop their leaves and stop growing each winter, with the slow-growing cell structure resulting in dense, long-lasting timbers. Hardwoods are often complex-shaped trees, such as oak, ash, elm and maple.

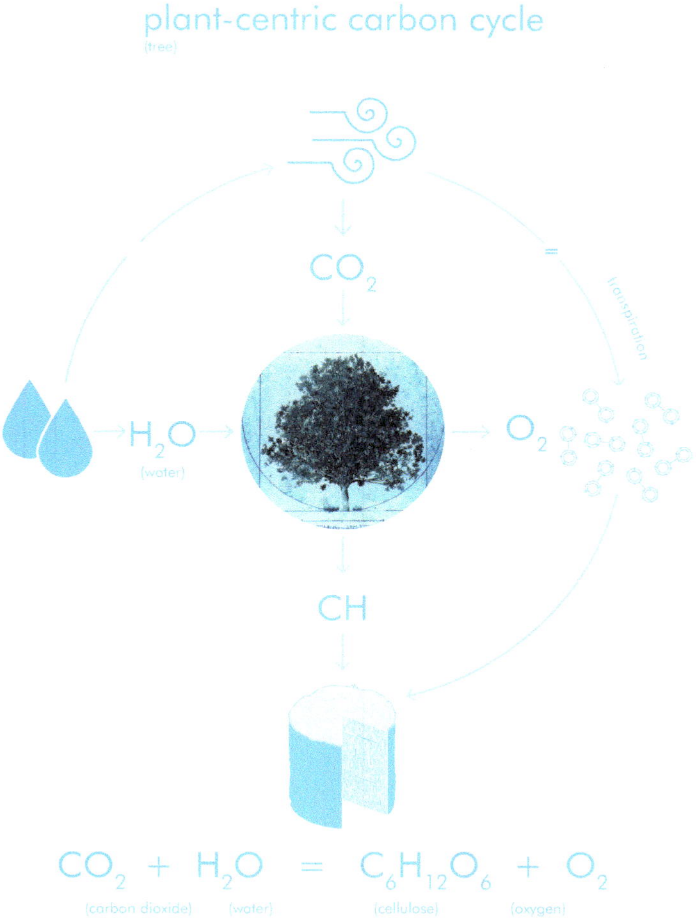

Figure 7.2 Plant-centric carbon cycle.
Source: Hayes/Marriage.

Softwoods are typically evergreen and coniferous, with thin needles for leaves that stay on the tree all year round. Their cell structure is frequently larger as it grows faster. Softwood conifers are species such as pine, spruce, larch and fir; trees that grow faster and straighter, often in forests that are further north (Canada and Siberia) or further south (Tasmania, Chile and New Zealand). It is primarily these straighter, faster-growing pine and spruce species that are used to manufacture the products of the future – the big blocks of timber that we call mass engineered timber.

Timber grain

As the cells of a tree are designed specifically to transport sap up a tree trunk from the roots up to the crown, the cell walls can also act the same way when a piece of cut timber gets wet. Water will get absorbed more easily into the end grain of timber than from the side, due to slicing across the cell structure. Exposed end-grain timber is therefore more vulnerable to weather. Wood is extremely porous and remains so after processing, with about 50%–60% of timber being open space between cell walls. With a moisture level of around 12% for timber when in use, construction needs to be detailed carefully to avoid moisture ingress which will cause swelling of the timber.

The strength of timber lies in the direction of the grain, with greater ability to withstand crushing loads applied directly parallel to the end-grain. Conversely, fixings into timber are always stronger across the grain rather than fixing into end-grain. Just as a tall tree will resist wind loads from the side when it is standing upright, we load timber from the same direction when it is cut and lying down. The spanning ability of timber is directly related to its depth and the width of the opening it spans (Figure 7.3).

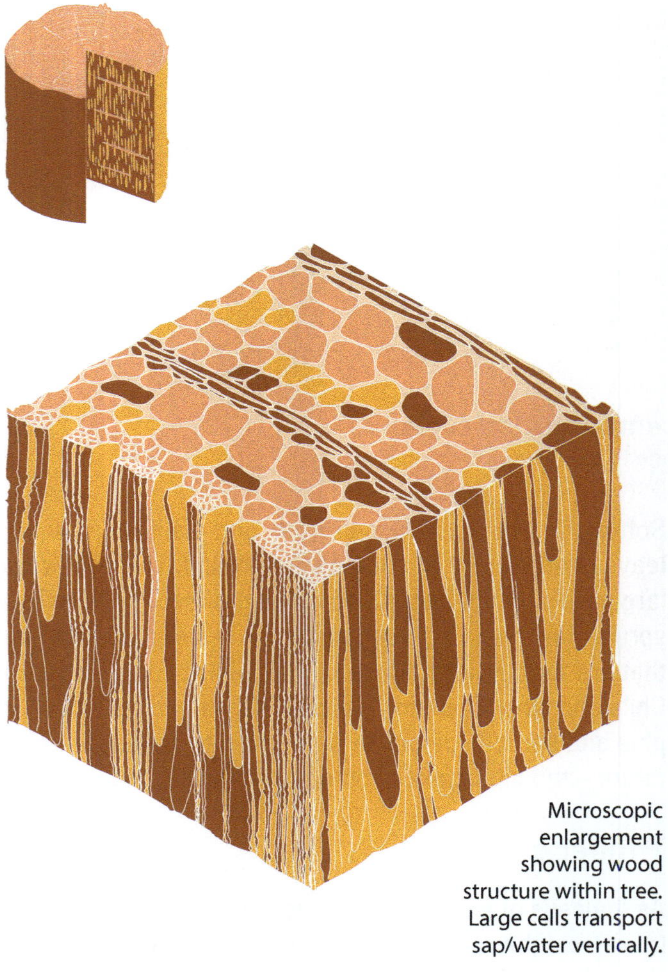

Figure 7.3 Cell structure within trees.
Source: Hayes/Marriage.

Traditional vernacular timber carpentry takes these factors into account, with woodworkers understanding the ability of which timbers could be used for mortice and tenon joints, how a scarfed timber connection works, or how to build in resistance to seismic shaking as evidenced by the highly skilled Japanese timber traditions of *shiguchi* (joining) and *kiwari* (dividing). Modern carpentry methods are cruder, usually consisting primarily of 90°cuts and an orthogonal system of bottom plate, studs, top plate, with intervening nogs or dwangs. In this traditional system, every stud must be contained by the top and bottom plates that spread the point load sideways, avoiding any point load that could imbalance the structure. In the world of 'modern means of construction' that is just around the corner, some of these materials and methods of connection are up-ended.

Modern means of mass engineered timber

Mass timber construction (MTC) is now a highly viable alternative to steel or reinforced concrete building construction, particularly in relation to residential buildings. In my previous book *Tall* (2019), we looked closely at the timber products available. For a fuller discussion on timber, I recommend that you read those *Tall* chapters. There are distinct differences in approach to design and construction between commercial office buildings and those buildings intended for residential use only. A brief overview of the many different types of engineered timber follows:

GLT

Glue-Laminated Timber (GLT or glulam) is made from small solid individual pieces of timber glued together under pressure and has traditionally been used for columns and beams in mass timber construction. Glulam has been around for longer than the other timber technologies noted here and has the unique advantage of being able to be easily formed into curved shapes. North American and European school sports halls such as ice-hockey rinks will often have roofs constructed over a grid of curved glulam beams. Because of the more hand-crafted nature of glulam, the product is more expensive than the more mass-produced LVL type products, but anywhere there is a need for a curved beam, glulam is really the only option.

LVL

Laminated Veneer Lumber (LVL) is composed of thin timber veneer sheets glued together, like a plywood, but with the grain of the timber all oriented in the same direction. This creates an incredibly strong timber beam capable of spanning considerable distance. LVL is likely to be used primarily as beams but may also be used as columns. In the case of columns, there may be a risk of crushing the end of timber fibres, so timber LVL columns may be fitted with steel 'boots' at each end of the column to disperse the structural load directly into the LVL without any crushing of the fibres.

CLT

Cross-Laminated Timber (CLT) is composed of solid pieces of timber (primarily softwood) securely glued together in layers, each layer being oriented at 90° to the previous layer. The production of CLT utilises numerous finger-jointed timbers, permitting defects such as knots to be cut out, so a homogenous softwood timber product is produced.

CLT is specifically designed for use as a two-way slab, capable of spanning reasonable distances according to its thickness. It can also be used as an integral part of the composition of timber shear walls. It is fabricated in an odd number of layers, often with three layers for wall construction. It should be noted that a three-layer composition will therefore have two thirds more spanning (and therefore load capability) in one direction than the other, and hence CLT is usually five layers thick for floor slab construction.

PSL

Parallel Strand Lumber (PSL) is another option for constructing timber columns, composed of ripped segments of timber laid in a matrix of phenolic glue, all oriented so that the grain of the timber parts is running in just one direction. The mixture is then set under heat and intense pressure, resulting in a dimensionally stable and very strong billet of timber – far stronger than a tree log alone. It also has the advantage of being able to be produced from off-cuts of other timber products, so makes good use of a resource that may otherwise have gone to waste. PSL is not suitable for use as slabs or beams, nor is it available in all markets, but it comes into its own as a material for mass timber columns. It has been successfully used for all the timber columns in the Brock Commons project at UBC in Canada.

DLT

Dowell-Laminated Timber (DLT) is composed of planks of timber that are held together with timber dowels, without glue. Usually the mass timber is a softwood with holes augered through several layers, while the hardwood dowel is hammered or screwed into place and connects them together. Particularly useful if your project is required to avoid the use of glues. Stiff connection is created by drying the dowels down to a low moisture content before inserting them into the slightly less dry base timbers: the dowel swells and locks tight as the humidity equalises.

NLT

Nail-Laminated Timber (NLT) is similar on build up to DLT but uses a steady pattern of nails to fix the timber together. NLT is for simple floor panels only and not suitable if you plan to do much milling of the timber, due to the risk of striking steel nails when machining. It is, however, usually a fairly cost-effective method of mass timber used for simple spans such as an internal mezzanine floor.

The future is timber

Each of these new timber products have their own specific advantage, but it is predominantly the use of CLT slabs that has unleashed new possibilities for apartment buildings. Invented by Gerhard Schickhofer in 1994 and developed in Austria over the past 30 years (Swedish Wood, 2020), CLT has been manufactured in only a select few countries so far but is now exported widely across the globe. Projects were confined to Austria for the first few years, for small dwellings like the Warrander Studio (Marriage and Sutherland, 2014), but more recently taller apartment blocks have established a whole new realm of construction possibilities.

The Murray Grove Apartments (Waugh Thistleton Architects, 2008) in London and the Forte Building (2009) in Melbourne were both innovative in assembling their buildings like a stack of giant cards – essentially a series of intersecting vertical CLT wall panels screwed directly together, rather than using the more traditional methods of platform framing or column/beam/infill construction. For a further, wide range of CLT projects, refer to Waugh Thistleton's *100 Projects UK CLT* (2018).

British Columbia (Canada) has been steadily pushing the envelope of what size building is permitted to be constructed out of mass engineered timber, pushing upwards with the Brock Commons building to 18 floors of timber, while Norway has been building even taller, with Treet and Mjøstårnet (refer Chapter 18) both reaching world record-breaking heights. On the converse side, the UK has only recently clamped down (post-Grenfell Tower enquiry) and is now stopping the use of mass timber in buildings taller than 18m. Many jurisdictions are still wary of this new technology, but those attitudes are constantly changing.

Building weight

When constructing apartment buildings from mass timber, there are some key advantages over other materials. Because mass timber is so much lighter than concrete (only one fifth the weight per volume) or steel (about one twentieth the weight), a mass timber apartment building will weigh far less than a concrete or steel-framed building. That has a beneficial effect on the cost, and foundation depth and complexity can be reduced substantially. Mass timber construction is considerably faster than concrete or steel/concrete composite structures (refer to Chapter 9). Mass timber also has advantages when it comes to fire resistance (refer to Chapter 12). The key thing to understand is the methods of connection.

LVL construction detailing

The jointing of timber columns and beams is distinctly different from the joining of steel beams and columns, or of concrete joint assemblies. Moment frame connections are much harder to achieve with timber beams and columns and necessitate the use of joiners such as steel plates or timber gussets.

The most basic joint has the timber column continue vertically through the connection point with steel T-spline brackets fixed to the column to pick up the timber beams. An alternative method features an engineered rigid steel node at the central joint with T-spline brackets welded directly to that. This allows for very strong mechanical (bolted) connections at the node, but note that this steel connection would need extensive fire protection as any weakening of the joint would have catastrophic consequences. Extensive coating of the steel elements with intumescent paint or encasing the entire joint with fire-resistance-rated plaster sheets are two options.

In heavily seismic areas, the joints need to be able to flex under seismic event loads without causing damage or collapse. Seismic systems developed out of Canterbury University (NZ) can work with either concrete or timber beams. The beams are held tightly onto the columns, but the joints can move at times of massive seismic events. Steel 'seismic fuse' technology can then absorb the energy, allowing the building to flex and then 'snap back' after the seismic event, with no damage. The PRESSS system is based on concrete beam technology (Concrete NZ), while PRES-LAM uses similar structural principles but is based on laminated timber technology (Pres-Lam.com). For advances with timber connections in an all-timber building, the Scion Building in Rotorua (NZ) is a world-leading example (Cattanach and Clarke, 2020).

CLT construction detailing

CLT slabs are an obvious, relatively straightforward solution to multi-unit apartment construction. CLT floor panels are laid out in an efficient pattern to achieve a regular two-way flat slab for building on. Floor panels are connected together with a simple spline joint from above or below and stitched together with screws. Pre-cut CLT wall panels are simply craned into place to form a standard rise (floor to floor) of panels on site, ready for the next set of floor panels. Repeat for the next layer, no beams are necessary. The millimetre accuracy of the computer-controlled cutting saws means that high-quality tolerance levels are achieved.

Manufacturers have extensive construction guides available online which detail the typical junctions utilised in CLT panel fixings (KLH and Stora-Enso). Perimeter (external wall) panels may now feature alternative edge profiles rather than the simple 90°edge cut as was used in Murray Grove, but the overall structural solution is still conceptually simple: create a rectilinear matrix of intersecting wall panels, firmly fixed together.

The SE will detail the structural needs of these junctions; most simply, the CLT slabs are stitched together with specialist, high-strength, thin-splined, self-countersinking, stainless steel screws developed by German and Austrian companies, that bite in and firmly hold these massive pieces of timber together. The mass of screws ensures that no movement is possible, aside from the natural flexibility of the timber itself. These screws will typically be either at 90° to the slab into the timber beam or wall below, or they can be installed at 45° when stitching a vertical wall slab to a horizontal floor slab. The SE will dictate the size and centres of the screws according to the static requirement (the analysis of the loads acting on the joint) but typically the screws will be at 100–150mm centres (4″–6″) continuously along the edges being joined. The screws are generally self-tapping and are easily installed with a battery-operated impact driver. The large number of screws evens out the stresses on the panels (and the screws) and means that no large bolts are required.

Another approach is to join the CLT elements together via small pressed steel angle brackets, most suitable for a wall to floor connection. These brackets and plates are fixed into place with numerous screws or nails, spreading the load out locally. Some of the brackets may be installed on the CLT floor slab prior to the CLT wall panel being craned into place, providing an accurate position guide to the CLT panel. Joint tape should also be laid on the join prior to the second panel being installed. Once the panels are assembled, any concrete screed topping layer to the CLT floor can be poured and this topping layer will encompass the steel brackets, sealing them within the floor sandwich.

Nail plates may also be utilised in certain locations, typically externally on the CLT joints. These can be installed to securely hold one wall slab onto another lower wall slab. Again, the plates are connected to the CLT with nails and the assembly is then covered with the external insulation and cladding. For further information on this last phase, refer to Chapter 13.

For very tall timber buildings, the edge load on the side of CLT panels should be avoided in case of crushing, so a combination of LVL columns and beams are interspersed with CLT floor panels. This approach has been undertaken on the two tallest residential timber towers in the world – at the time of writing, both in Norway.

Brock Commons Tallwood House (refer Chapter 18) uses a similar steel node connection detail, although in a fairly radical move, this structure has no beams – the timber CLT slabs sit directly on the steel nodule atop the timber PSL columns. The steel connection node is completely enclosed within the CLT floor slab and a concrete topping slab encompasses the steel joint.

CLT shear walls

CLT can be used in a shear wall, often in conjunction with LVL billets. The dimensions of the CLT panels commonly vary from 60mm to 350mm thick in three, five, seven or nine layers depending on the strength required to resist loads. As the panels are designed and manufactured off-site, their sizes are mainly dependent on the transport capacity and are commonly around 3.5m × 15m. Depending on the required resistance and ease of assembly on site, panels can be either installed vertically, cantilevering several storeys high, or installed floor by floor. In the first case, a balloon-frame design approach is used where floors rest on metal brackets or a corbel, and are then fixed to the wall using nails or screws. In the second option, similar to the platform method, CLT panels stop at each floor and floor panels rest directly on top of them. A range of different types of fixing can be used from the simplest – long wood screws – to a more complex design using threaded rods which join together the top and bottom vertical panels via the floor slab. Whichever joining system is used, the strong connections ensure that the CLT floor slabs and walls work together as a combination of shear walls and diaphragms.

There is much to still be explored on the use of CLT shear walls in the design of high-rise buildings, and their potential impact on the planning layout. As solid panels, CLT helps establish clear and definitive boundaries which can be particularly relevant when designing residential buildings. The apartments' layout would be directly defined by the building's main structure but the risk is to reach an excessive spatial compartmentalisation with highly enclosed spaces. A solution is the use of the rigid frame made out of glulam or LVL as a second structural system, in order to maintain a full timber design. This would help the building reach greater heights while providing wider planning flexibility for apartment design. For an example of this, refer to the Case Study Chapter 19.

In the next chapter, Emily Newmarch examines the importance of the embodied carbon inherently in our timber.

Works cited

Cattanach, Alistair and Clarke, Ryan. 2020. "Scion hub: a three storey seismically resilient timber diagrid." *NZSEE 2020 Annual Conference*. Wellington: New Zealand Society for Earthquake Engineering.

Concrete NZ. *PRESSS Design Handbook*. Available at: https://www.concretesociety.org.nz/index.php/members/recent-seminar-technical-publications/viewdownload/5-seminar-technical-publications/12-presss-design-handbook

KLH, n.d. *Component Catalogue for Cross Laminated Timber Structures.* Austria: KLH Massivholz GMBH. Available at: www.klh.at/en/klh-construction-en.pdf

Marriage, Guy. 2019. *Tall: The Design and Construction of High-Rise Architecture.* London: Routledge.

Marriage, Guy and Sutherland, Ben. 2014. *New Digital Housing Typologies: CNC Fabrications of CLT Structure and BIM Cladding.* Architectural Science Association: ANZASCA.

Pres-lam. Available at: https://pres-lam.com

Stora Enso. 2020. *Massive Wood Construction.* Available at: https://www.storaenso.com/en/products/wood-products/massive-wood-construction

Swedish Wood Magazine. 2020. "A global solution for a locally active industry." *Wood Magazine*, issue 2, 2020. Available at: https://www.swedishwood.com/publications/wood-magazine/2020-2/gerhard-schickhofer/

Waugh Thistleton Architects. 2018. *100 Projects UK CLT.* London: Softwood Lumber Board and Forestry Innovation Investment + Waugh Thistleton Architects.

08

Chapter Eight

Embodied carbon

Emily Newmarch

Embodied carbon

DOI: 10.4324/9781003123873-8

Introduction

There is a strong and urgent need to create more houses for people and to reduce the amount of land being used, much of that new housing will need to be in the form of apartments. If we continue to make that new housing out of our traditional range of materials, however, we will run into even more issues. The concrete we specify, the steel beams and columns, the glazing, the aluminium: all are contributing to ongoing problems. Problems don't just occur while we use the building, but long before and long after. Architects, developers and contractors need to weigh up the impact of their decisions and give careful consideration to the materials that cause the least amount of harm to the environment.

Why embodied carbon matters

Carbon is one of those pesky little atoms that likes to bond with many other atoms, including more of itself. Its ability to bond easily and create a wide range of stable structures means that it can be found in so many different things, including building materials. Carbon becomes a big problem when it bonds with oxygen and creates carbon dioxide – that heavily reported gas that scientists have been warning us about for decades. For building materials, carbon dioxide is emitted, one way or another, at every point you do something to that material which is outside its natural cycle. This includes extracting and processing the raw materials, transporting the materials, putting them together on site, maintaining them, replacing them, and eventually disposing of them. The impact of each step is recorded as embodied carbon. A positive number indicates that an excess has been released into the atmosphere, and a negative number indicates that carbon is being stored within the material. In the context of global warming, a positive number is contributing to the problem and a negative number is mitigating it.

Operational carbon

Measured using the same units, operational carbon is associated with the way we use buildings. Reducing operational carbon heavily benefits from the large body of knowledge that exists under energy reduction.

Keeping track of carbon emissions – life cycle assessment (LCA)

Keeping track of how much carbon dioxide is released into the atmosphere is a key step in becoming more aware about the impact a building has on the environment. Carbon emissions, embodied or operational, are recorded as a buildings Global Warming Potential (GWP), which represents carbon dioxide or equivalent greenhouse gas in kilograms ($kgCO_2$-eq). However, measuring a gas isn't something we do every day. It

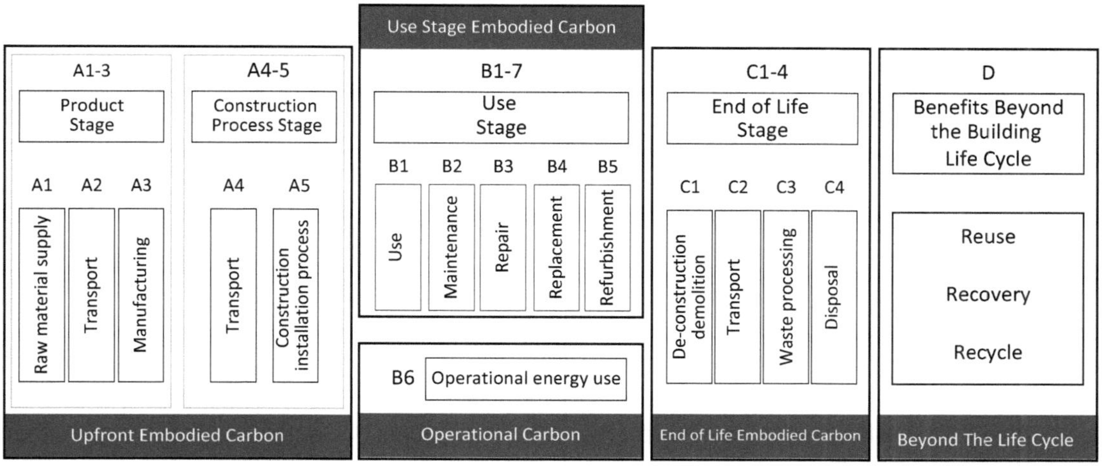

Figure 8.1 Embodied carbon life-cycle module diagram.
Source: Newmarch.

can feel like a complex task that is difficult to relate to. Following a clear set of rules breaks down the task of measuring carbon emissions and makes it easier to compare results. This is where Life Cycle Assessment (LCA) comes in. There are a few standards that set rules for how an LCA should be done for any type of product. For buildings, EN 15978 outlines a simple set of modules that creates categories to record emissions at different life cycle stages (Figure 8.1). Many tools that are available on the market reference this standard when talking about their scope. Scope is important, as getting a full picture of all the emissions associated with a building's different life cycle stages will help guide more informed decisions.

Assessing GWP from embodied carbon can also be put into context with the financial cost of building materials through a combination of marginal abatement cost curves and carbon cost analysis, an approach developed by Sturgis (2017). Adding additional layers of information alongside an LCA early in the design process can help set up a project for continued efficient analysis. An LCA can also contribute to achieving accreditation systems such as Greenstar, LEED, BREEAM and more.

Current challenges

To date, there has been an enormous wealth of information surrounding the way we operate buildings. We can see evidence of this knowledge in the number of building codes that regulate minimum building performance, and the many accreditation systems that set the benchmarks for 'best practice' and 'state-of-the art' performance. However, the intense focus on the way we use buildings has drawn attention away from the building itself. For a long period, the concern about the rising cost of energy has left the rising cost to our planet on the sideline. We are challenged with a sizeable gap in everyone's knowledge on how best to reduce our emissions and how to do it quickly. In many countries, this gap is present in the lack of minimum regulation to enforce limits on how much embodied carbon can be emitted from building materials.

In the act of measuring embodied carbon, one of the biggest challenges is getting data on materials that are reliable and appropriate for the country they are applying to. Two products may have a similar appearance but could have different manufacturing processes, be produced in different parts of the world, have different service lives and may or may not be reusable or recyclable at the end of its life. Because of the wide variation being presented in multiple different life cycle stages, it is hard to trust embodied carbon material data sourced from generic databases. Whilst it is better to have a rough idea than none, there is a huge need for product companies to become more transparent about the carbon emissions associated with each life cycle stage.

Further, the negative effects of global warming that building practices contribute to, cause a grey cloud to float over the idea of analysing embodied carbon. Confronting the issue can bring up feelings of guilt, which only leads to withdrawal and putting distance between a project and the problem. The grey cloud of negative consequences dominates any engagement and can be challenging to overcome. Talking positively and optimistically about creating low carbon buildings opens an opportunity for innovative leaders all over the world to inspire others. Part of the reason embodied carbon is so difficult to understand relates to the fact there are multiple pathways to solve the problem. Having multiple options leaves a lot of room for different design styles to be integrated into a solution.

Environmental product declarations (EPD)

When exploring options around material selection, a good thing to look out for are Environmental Product Declarations (EPD). An EPD is an independently verified and registered document that details the environmental impact of a product. The GWP will be listed along with many other environmental impacts that can be used to deepen a comparative analysis. Producing an EPD is not a cheap process. The companies who make the effort to be transparent about the good and the bad, demonstrate a willingness to learn about their own process and how to improve it. For the companies without an EPD, it is important for designers to question them about their manufacturing processes and end-of-life plan. These can be little questions like; where do you source the raw material from? How and where do you manufacture your product? Can the product be recycled or reused? These little questions from people within the industry can have a big impact in demonstrating there is a demand and a need in having quantifiable answers to these types of questions to keep a product competitive.

Are modern apartment buildings cursed?

Much of the existing research into embodied carbon for large-scale buildings has used office buildings as case studies. A key point of difference between an office and an apartment building is the density of each floor plate. An office building is typically very open, with only essential dividing walls for tenancies or toilets. On the other hand, a modern apartment building is far denser, condensing as many functional rooms into each floor plate as possible. The result is a larger amount of materials being used to create these divisions, and extra servicing needs like kitchens and bathrooms. However, the larger quantity of materials required for modern apartment buildings isn't a curse that implies a much higher GWP from emissions associated with embodied carbon. Instead, modern apartment buildings can have one of the biggest advantages in achieving embodied carbon reduction targets.

The potential for large reductions is in all the additional divisions, kitchens and bathrooms within modern apartment buildings. These elements have more flexibility around material selection as they are not bound by preforming structural or weather tightness duties like the primary structure or building envelope. Here, an easy win would be to specify materials that have net negative value, meaning that they would store more carbon than would be emitted to produce it. An obvious choice is wood grown in a sustainable forest and manufactured locally. Wood can be used as the structure of the divisions or as an internal finish. The benefit from using a negative embodied carbon material in large quantities is the balancing effect it has on the overall net impact of the building. For example, if the structure and envelope of the building had an embodied carbon value of 1,500 kg CO_2-eq/m^2 and the interior divisions and finishes had a value of –1,000 kg CO_2-eq/m^2, the net result for the entire building would be 500 kg CO_2-eq/m^2. Having the ability to still create balance when big wins, like structure or envelope, are not economically viable, provides flexibility and opportunity for modern apartment buildings.

Arguing about sustainability

Clearly the world needs to build more housing, to solve the chronic housing shortages growing all over the world. Clearly too, however, the construction industry needs to build more efficiently and use more low carbon or zero carbon methods for doing so. There is no magic material that can save the day when it comes to mitigating the effects of global warming. However, understanding the arguments for and against key material groups can help support smarter and more conscious use of materials.

Wood

Many governments around the world have adopted what are known as 'Wood First' policies. There are many political reasons for adopting such policies that vary from country to country, but key amongst them is the reason that wood absorbs and sequesters carbon dioxide and hence helps combat climate change. The vast excess of CO_2 being produced by the world needs to be captured and one of the most helpful, naturally occurring ways of doing this is to encourage the growing of trees and the use of mass timber in long-life buildings.

Whilst trees will play a significant role in slowing the effects of global warming, they are not a bullet proof solution. Large forests are still vulnerable to drought, diseases and wildfire. Having large single-species areas also isn't healthy for the soil quality. Sustainable forest and land management legislation provides reassurance that the market demand for wood doesn't get taken advantage of. When specifying wood products, always look for verification that the product has been produced sustainably before assuming the product will have a big carbon sequestration ability (Figure 8.2).

Steel

Marketing for a variety of products often focuses on the circularity potential of steel. While steel is very capable of being a part of a closed loop recycling system, there is a large upfront embodied carbon cost to producing the material. In an EPD, the end of life benefit is recorded as a negative under module D, slightly balancing out the high positive values recorded under modules A1–A3. The module D benefit is dependent on the product being recovered at the end of a building's life, which can be difficult to promise. To ensure this a possibility, the design must enable the recovery of the material. In pure steel structures, this is usually possible. However, recovery becomes difficult when the steel is buried within concrete, which requires a large amount of energy to decontaminate the raw material.

Comparing steel to wood requires an understanding of time. Both products have arguable sustainable benefits, they just occur at different life cycle stages. Wood provides a highly attractive negative embodied carbon value at the product stage, but can be easily contaminated during the life cycle of the building, turning the product into waste at the end of its life. Steel, on the other hand, requires a large embodied carbon cost to convert the raw material into a viable construction material, but can be recovered and reproduced at the end of the buildings life. The embodied carbon cost to recover and reproduce a steel product is less than the

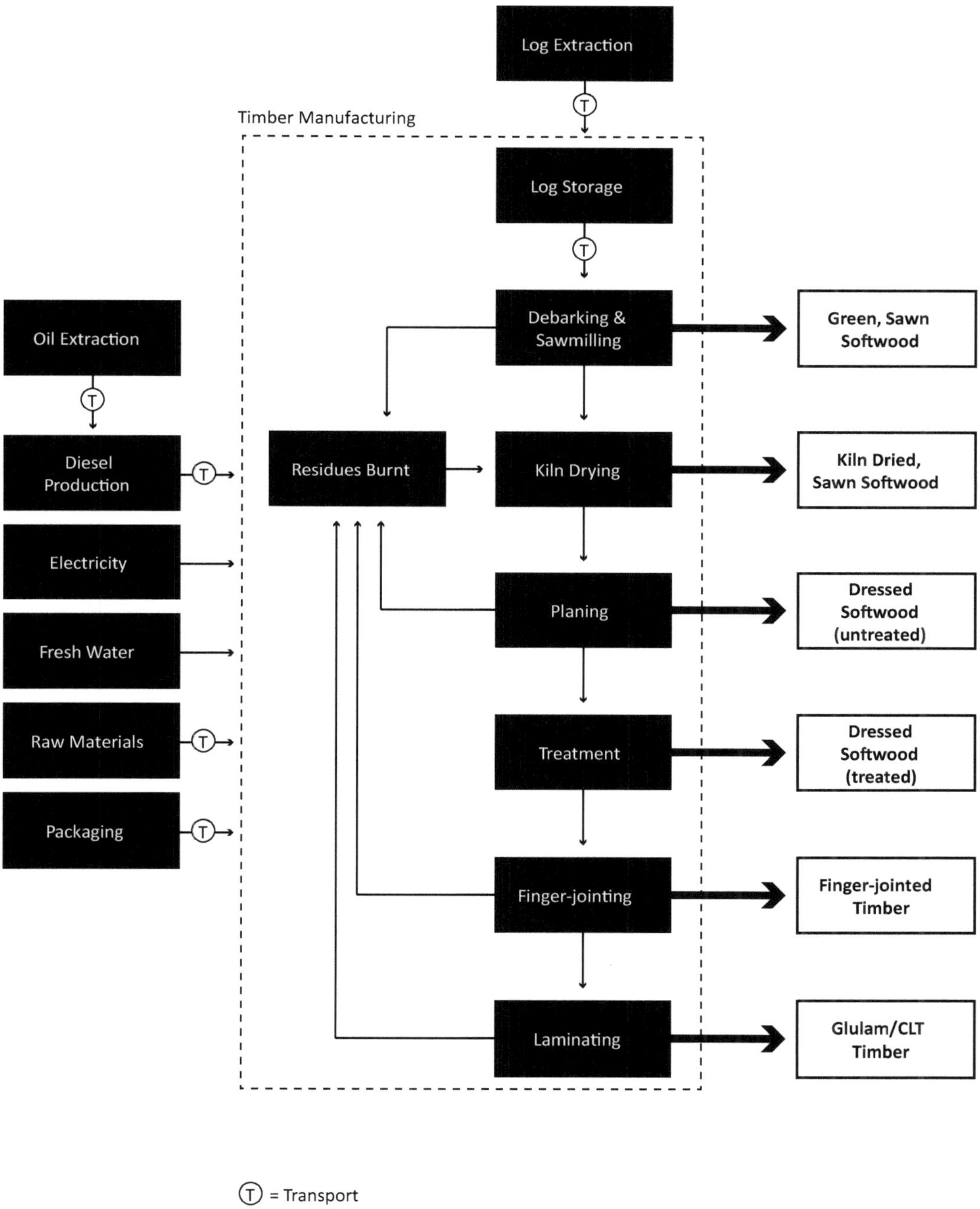

Figure 8.2 Embodied carbon flowchart of timber manufacturing.
Source: Newmarch.

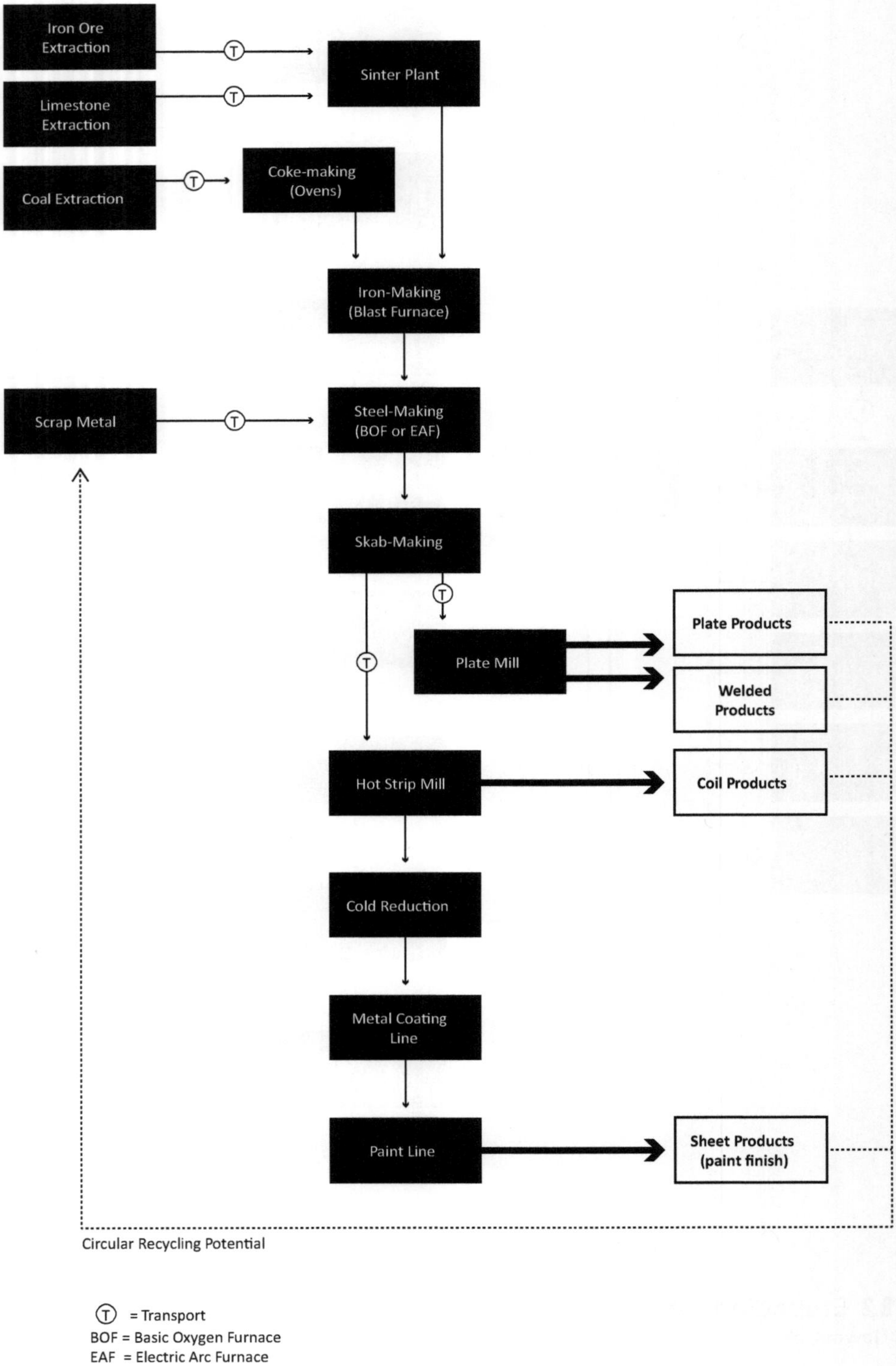

Figure 8.3 Embodied carbon flowchart of steel manufacturing.
Source: Newmarch.

cost to begin from raw iron ore, and hence is presented as a negative saving (Figure 8.3).

Concrete

Concrete is like cake mix, there are many ingredients that come together to produce the final product. A standard concrete mix will have a high embodied carbon cost due to the use of cement as a core ingredient. Cement has a large footprint due to being heated at extremely high temperatures to produce clinker. The production of clinker can account for 90% of the total embodied carbon footprint. Due to the high carbon cost of a single ingredient, many substitutes have emerged on the market. If concrete is a necessary component on a design, talking to your local manufacture about what cement alternatives are available to your industry is a good way to reduce the impact of a stereotypically 'bad' material.

Concrete isn't always 'bad' all the time. A common argument for concrete as a sustainable material comes back to the material's ability to outlast the lifespan of many other materials. This sounds great but it limits any renovation possibility if the building changes uses. Again, the comparison to other materials comes down to a conversation about time. In many countries, there are targets like net zero by 2050 which respond to the fact there is a limit on how much carbon that can be emitted before we hit the 1.5°C of warming milestone. This milestone was set in 2018 by the IPCC after further research revealed the damage the 2°C would bring that was set in 2015 (IPCC, 2018; United Nations, 2015). These global limits and national target reinforce that the timing of our emissions is critical. Specifying large upfront emissions with the potential to recover at the end like steel, or materials that will last forever like concrete, is not a viable argument for every new building (Figure 8.4).

Strategies for designing modern apartments

The list below has been adapted to the context of apartment buildings from general advice in the *Climate Emergency Design Guide* by LETI and the *Bringing Embodied Carbon Upfront guide* by the WGBC (Adams et al., 2019; LETI, 2020). See Table 8.1.

Future growth

No matter what the future holds for us, every option that we choose has an element of carbon associated with it. Whilst there is no 'one-size-fits-all' answer that can be applied across the globe, there is hope in the potential for diversity and creativity in finding solutions. Developing hybrid systems that introduce timber to balance out the positive and negative emissions from other materials, like steel and concrete, can help any project achieve

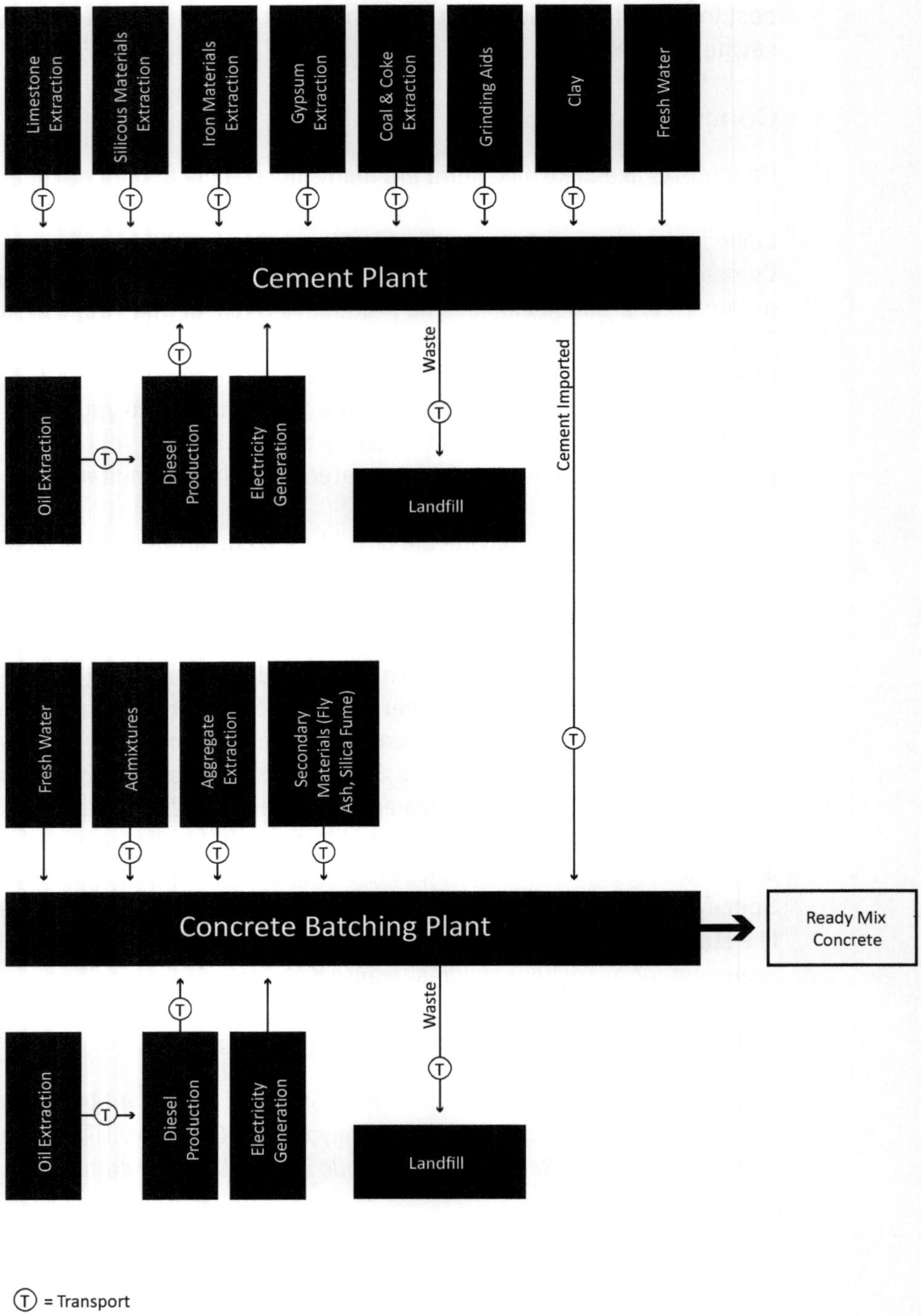

Figure 8.4 Embodied carbon flowchart of concrete manufacturing.
Source: Newmarch.

Table 8.1 Strategies for designing modern apartments

Strategy	**Consideration**
Build nothing	Is it necessary to construct a new building? Can an existing building be converted into apartments instead?
Build less	If there is an existing building onsite can part of it be reused? Can the structure be retained and strengthened to increase its life? Consider using lighter structural systems to keep the weight of the building low. Where possible, avoid creating large voids that require large spanning structural elements.
Build smarter	Reduce material waste by integrating standard sizes into the design. Look for efficient construction practices like prefabrication. Consider repeatable modules at different scales like bathrooms or an apartment unit.
Build consciously	Be aware of the service life of materials and favour products that will last longer and require less maintenance. Look for low embodied carbon alternatives. What materials are available locally?
Build beyond	Design to allow easy reconfiguration of spaces as the use of the building may need to change in the future. Think of the building as a material bank, where materials can be reused or recycled at the end of the apartment building's life. Opt for mechanically fixing over glues or adhesives. Avoid materials that contaminate other materials. Have an end of life plan for the building.

Source: Adapted from Adams et al. (2019) and LETI (2020).

big wins in the structure and envelope building elements. For modern apartment buildings, continuing to shift the net carbon balance of a project by sequestering carbon within the interior divisions and finishes further helps the building typology move forward with a smaller footprint.

Works cited

Adams, Matthew, Burrows, Victoria, and Richardson, Stephen. 2019. 'Bringing embodied carbon upfront: Coordinated action for the building and construction sector to tackle embodied carbon'. Report. London, United Kingdom: World Green Building Council.

IPCC. 2018. Global warming of 1.5°C." Special Report.

LETI. 2020. *LETI Climate Emergency Design Guide*. London, United Kingdom: London Energy Transformation Initiative.

Sturgis, Simon. 2017. *Targeting Zero: Embodied and Whole Life Carbon Explained.* Newcastle upon Tyne: RIBA Publishing.

United Nations. 2015. "Paris Agreement."

09

Chapter Nine

Prefabrication

Prefabrication

Pamela Bell

DOI: 10.4324/9781003123873-9

WHAT do we call it?

There are many terms for prefabrication. These terms are used interchangeably, and confusingly, by different people in the construction sector depending on their historical perception or personal preference. But no matter what term you use, you are effectively talking about any work that is performed BEFORE reaching the construction site. Here are some of those terms – pick your term and stick with it, but stay open to substitution:

- Prefabrication
- Prefab
- Innovative construction
- Offsite or off-site
- Offsite construction
- Offsite manufacture or OSM
- Modern methods of construction or MMC (despite many methods not being particularly modern or recent at all)
- Transportable (although technically this is just one type of prefab)
- Pre-designed
- Pre-built
- Pre-manufactured
- Pre-assembled
- Pre-manufactured Volumetric Construction (PMVC)
- Mass customisation
- Standardisation

Historical misperceptions in the United Kingdom have moved the construction culture away from the term 'prefab' and more towards 'offsite'. A revitalisation of 'modern green prefab' in the West Coast of the United States of America has made the prefab term more relevant there (Bell and Southcombe, 2010). The Southeast Asian construction industry focus on affordable concrete boxes has meant a preference for volumetric terms such as PMVC.

Prefabricated typology framework

Any system needs a framework, and at its simplest, methods of prefabrication can be broken down into five key types. In New Zealand, the five-typology framework developed and successfully promoted by Bell is used as a clear guide to the industry by the association PrefabNZ (Bell, 2009). These five types are from smallest to largest (Figure 9.1):

- Component
- Panel

1

Components are relatively small scale items that are invariably assembled offsite, such as light fittings, windows, and door furniture. It includes structural members (trusses and frames), fittings, fixtures, and joinery that is cut, sized or shaped away from the site for assembly on site.

2

Panels are planar units that do not enclose usable space, such as panel systems and cladding panels. They may include windows, doors or integrated services, and are either open-framing or closed-in with cladding and/or lining. They are transported to site as flat-packs.

3

Volumetric prefabricated units enclose usable space and are then installed within or onto a building or structure. They are typically fully finished internally, such as toilet/ bathroom pods or plant-rooms. Refers also to pod.

4

Hybrid-based prefabrication is also referred to as semivolumetric prefabrication. It consists of a mixture of volumetric or modular units and non-volumetric or panelised units (module plus panel). It may also include component and site-built elements.

5

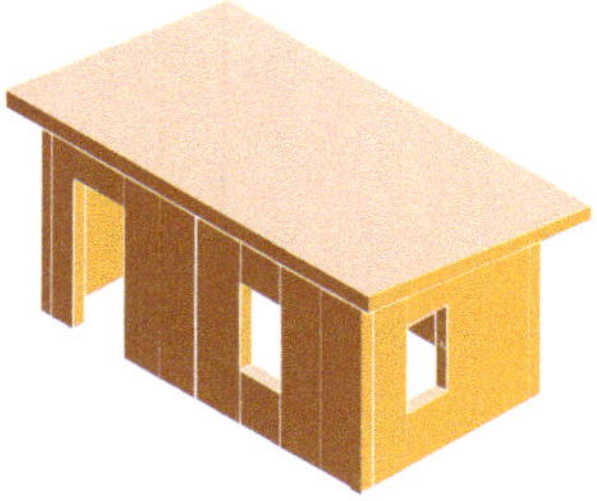

Complete buildings are fully-finished volumetric construction. It is typically yard or factory finished internally (and possibly also externally), to make a single building or parts of a multi-unit building. In New Zealand, complete building prefabrication includes the traditional transportable housing industry.

Figure 9.1 Five categories of Prefab.
Source: Hayes/Irvine, after PrefabNZ.

- Volume
- Hybrid
- Complete building or transportable

In the United Kingdom, a Modern Methods of Construction (MMC) definition framework has been developed that outlines seven categories of pre-manufacturing (MHCLG). Far more complicated and harder to define/assign than the PrefabNZ system, the seven categories outlined by MHCLG are:

1. **Pre-manufacturing** (3D primary structural systems) – includes pods, volumes, modules or mods
2. **Pre-manufacturing** (2D primary structural systems) – includes panels, open frames and trusses
3. **Pre-manufacturing components** (non-systemised primary structure) – includes other smaller pieces and parts that combine to create panels or larger structural components
4. **Additive manufacturing** (structural and non-structural) – includes 3D printing
5. **Pre-manufacturing** (non-structural assemblies and sub-assemblies) – includes staircases, cabinetry and joinery
6. **Traditional building product led** site labour reduction/productivity improvements – includes robotic automation used at site for activities such as brick-laying
7. **Site process led** site labour reduction/productivity/assurance improvements – includes robotic production of pieces and parts, as well as software technology applications

WHAT are the basic principles or WHY do we use it?

In the United States, Professor Ryan E Smith is an established academic authority on all things prefab or modular or offsite, being Professor in the School of Design and Construction at *Washington State University*. In *Prefab Architecture: A Guide to Modular Design and Construction* (2010), he outlines six key principles of construction that affect productivity and provide reason to explore prefabrication as a potential contributory solution (p. 80):

- **Cost**: capital and operational investment
- **Labour**: skilled and unskilled human workforce
- **Time**: schedule or duration of the project
- **Scope**: extent or breadth of the project
- **Quality**: design and construction excellence
- **Risk**: exposure to potential financial loss

Importantly, Smith also outlines the hidden costs in using prefab (p. 83):

- **Overhead**: manufacturing facilities, equipment, maintenance, etc.
- **Profit**: needed to charge in order to cover overheads above, which may bring overall cost in line with traditional construction
- **Transport**: may be higher than traditional due to necessity to ship with 'air' once premade into pieces and parts, rather than flat-packed
- **Setting**: lifting, craning and manoeuvring can be awkward and demand specialist skills
- **Design fees**: more coordination is needed earlier in the project which means more investment in the design team up-front (Figure 9.2).

If you are still uncertain whether your project is suitable for prefab or not, a helpful way to assess the 'why' is by comparing these basic principles in further depth. Tables 9.1–9.6 are designed to assist architects in determining the advantages and disadvantages of using prefab when planning a new project (Smith, 2010, 95–97).

Key points about evaluating cost-based principles when exploring prefabrication for your project:

- Remember that saving time is the fastest way to save money – time is a major driver of value for many construction project clients (see the Time table below)
- Intelligent quantity surveying cost analysis will be needed to transfer time savings into visible bottom-line savings for clients
- Growing industry awareness for calculating lifecycle costings and productivity improvements as sustainability concerns become more paramount (Figure 9.3)

Key points about evaluating labour-based principles when exploring prefabrication for your project:

- Prefabrication uses computer-controlled technology and indoor working environments that benefit workers in a number of ways
- Working indoors at grade, instead of at height, provides a safer environment
- Workers are likely to develop a broader range of light-touch skills in a controlled environment, rather than a deep niche skill-set as at site.

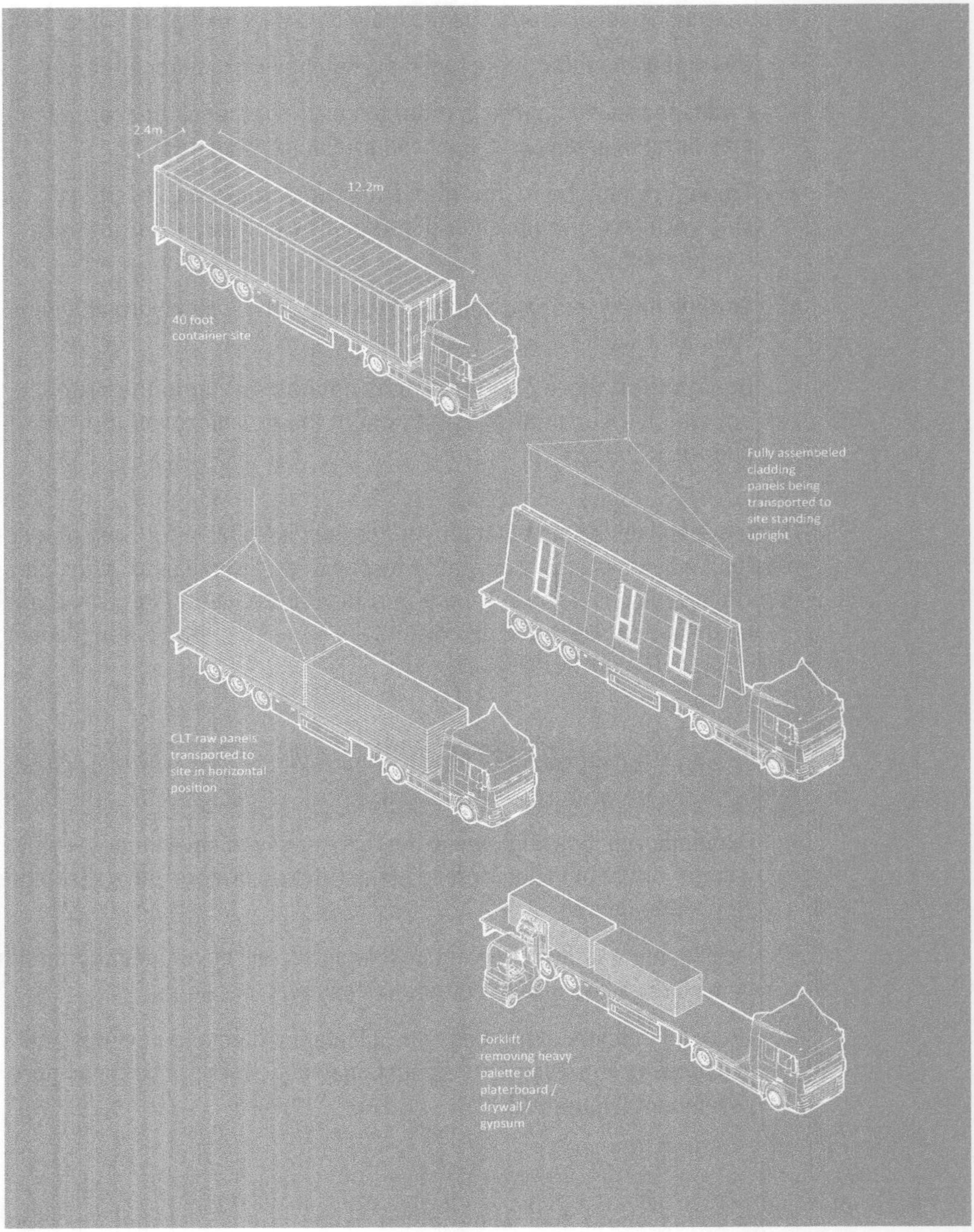

Figure 9.2 Transportation and lifting.
Source: Hayes/Marriage.

Table 9.1 Cost: offsite versus onsite

Principles	Offsite	Onsite
Cost		
Financing	Interest reduced on shortened schedule, even draws, and leasing options, alternative methods might be seen as risky for lenders	Traditional construction loan/mortgage financing, Lending freezes make construction actuation difficult
Administration	Administrator overhead reductions	Bureaucratic layers for decision making
Insurances	Lower contingency costs	Higher contingency costs
Transportation	Two-staged delivery shop and site	Raw material delivery only
Change orders	Extra cost and delay	Accommodated changes
Overhead	Larger shop overhead – people, equipment, space, utilities	Overheads absorbed into construction budget
Schedule	Duration reductions recapture investment earlier	Schedule overruns are common, increasing overall budget
Material	Less scaffolding formwork and shuttering	Increased scaffolding formwork and shuttering
Craning	Costly heavy-duty cranes for setting	No cranes for small projects large stationary crane for larger
Initial cost	Higher investment in product	Lower initial cost for normative projects
Lifecycle cost	Greater ROI over long term	Greater maintenance requirement
Profit	Subcontractor overhead costs project more, savings from scope, material may not be passed onto customer	Overhead fees are more transparent to owner
Design fees	Higher due to coordination requirement	Standard fees
Lean	Reduce time waste, increase value	Waste laden process
Productivity	Full 8 hours of work, sophisticated machines, digital tools available	Productivity increases difficult
Economy	When strong, plenty of residential work, but less commercial; when weak, less residential and more commercial	Residential and commercial ebb and flow with markets

Based on Smith (2010). Used with permission.

Table 9.2 Schedule: offsite versus onsite

Principles	Offsite	Onsite
Schedule		
Duration	Finish date, met 50% reductions	Schedule overruns are common
Scope coordination	Extra coordination needed between site and plan	More time for coordination and opportunity to adjust dimensions
Schedule reliability	Longer lead time, reduced erection time, reliable duration	Shorter lead time, longer construction and less reliable
Permitting	Streamlined in familiar jurisdictions, opposite in unfamiliar	Dependent on jurisdiction
Weather	Sun always shines	Delays due to weather are common
Work flow	Concurrent scheduling	Linear process
Subcontractors	Fewer conflicts, better sequencing	Simultaneous trade, crowding difficult
Supply chain management	Coordinated and streamlined	Uncoordinated and wasteful

Based on Smith (2010). Used with permission.

Key points about evaluating time-based principles when exploring prefabrication for your project:

- As mentioned earlier, saving time is the fastest way to save money – time is a major driver of value for many construction project clients
- Saving time is of most interest to large-scale accommodation providers such as student accommodation, retirement villages, apartment projects, hotels, etc.
- More time will be needed for design coordination prior to site works – as all design decisions must be made and frozen prior to manufacture of offsite elements
- Staged timing of delivery of large elements to reduce storage time at site is critical – otherwise space at site (or in a nearby facility) will need to be arranged

Key points about evaluating scope-based principles when exploring prefabrication for your project:

- Coordination throughout a prefabrication project is critical – you want to have your 'dream team' around you on day one

Table 9.3 Labour: offsite versus onsite

Principles	Offsite	Onsite
Labour		
Local labour	Less local labour needed	Local labour needed
Working conditions	Improved working conditions and more stable job market	Variable working conditions and more sporadic job market
Skill level	Craft and technical skills needed	Craft and problem skills are elevated
Subcontractors	Fewer conflicts, better sequencing	–
Unskilled labour	Supervision of labour, quality control process	Unsupervised labour leads to portions of project being reconstructed
Labour comfort	Ergonomics increased	Physically difficult
Safety	Reduced exposure to accident	Accident prone job site
Health	Better life style and mental health	More opportunity for variety in health
Skilled labour	Less chance for skill development	More chances for skill development
Commute	Factory near house – full 8-hour days and no out of town travel	Out of town projects require commute times
Productivity	Full 8 hours of work, sophisticated machines, digital tools available	Less productive use of labour force
Union	Declining due to immigrant population making less room for offsite	Accommodates variety of labour types

Based on Smith (2010). Used with permission.

- Early engagement with design, manufacture and contracting is important to ensure the buildability of the design
- Complete design package with frozen design and shop drawings for manufacture means greater emphasis on this part of the process than in traditional projects (Figure 9.4).

Key points about evaluating quality-based principles when exploring prefabrication for your project:

- Quality is value-based and subjective according to each client – most are wanting a reliable product delivered on time and on budget which favours the use of prefabrication
- Manufacture of a repetitive part offsite means prototyping and tight quality control to achieve expected standards

Table 9.4 Scope: offsite versus onsite

Principles	Offsite	Onsite
Scope		
Supply chain management	Long-term supply chains for materials established	Supplies restricted to project-based purchases
Coordination	Extra coordination needed between site and plan	More time for coordination and opportunity to adjust dimensions
Flexibility	Changes often cannot easily be made in field	Limited adjustment can be made easily in the field
Impact of changes	Less accommodation	More accommodation
Maintenance	Reduced maintenance and operations	Higher maintenance and operations
Transportation	Two-stage delivery shop and site	Raw material deliver only
Flexibility	Changes not made in field	Adjustments made in field
Design	Requires higher level of detailing for assembly, only 50% with bridging documents	Design intention communicated only
Production	Predictable output, mockup, and prototype required	Difficult to anticipate, depends on skill level of construction crew
Regulatory	Third party verifiers	Local agency to inspect
Predictably	Increase expected outcome	Less predictable delivery
Staging	Less material on site, but must be coordinated well	Staging is logistically difficult
Accessibility	Specialised companies, takes research and work	Smaller construction companies

Based on Smith (2010). Used with permission.

- Levels of innovation and sustainability benefits are achievable in the controlled setting of an indoor environment where waste capture and reuse are easier than in an outdoor site (Figure 9.5).

Key points about evaluating risk-based principles when exploring pre-fabrication for your project:

- Each project and each client has their own value basis and risk tolerances which need to be established alongside evaluating offsite versus onsite delivery options
- Both qualitative and quantitative measures need to be taken into account when evaluating costs, benefits and risks (Figure 9.6).

Table 9.5 Quality: offsite versus onsite

Principles	Offsite	Onsite
Quality		
Reliability	More reliable quality can be achieved in shorter amount of time	Less reliable (depending on the site conditions and skill level of labour)
Coordination	Integrated effort between factory and site	Flexible coordination and adjustments
Design	Integrated design and construction process	Separation of design and construction
Production	Predictable output, mockup and prototype required	Difficult to anticipate, depends on skill level of construction crew
Regulatory	Third-party verifiers with industry knowledge	Local jurisdiction with varied experience
Predictably	Increase expected outcome	Unpredictable quality
Innovation	R&D capacity and control	No research and development time or resources
Design flexibility	More restricted	More freedom
Equipment	Easier access	Equipment to and from site
Environment	Lower waste, air and water pollution, dust and noise, and overall energy costs	Difficult to manage waste and energy in construction
Handling	Potential for damage during handling	Smaller elements easier to handle
Joining	Fewer joints, but difficult to detail	More joints, more potential for failure
Tolerances	Great capacity, not forgiveness in module on site	Forgiveness with details constructed on site
Fit	Fewer points for water and air infiltration	More locations for infiltration
Quality of materials	Quality control in SCM sourcing	Contingent upon source
Warranty	Opportunity for comprehensive warranty of products from one supplier	Dedicated to each system supplier

Based on Smith (2010). Used with permission.

WHERE and WHEN to use prefab in your project

Once your 'why' for using prefab is established, the next step is to work out which system to use, when and where. The 'when' is easy – employ a prefabrication approach as early as possible. The 'who' is easy too – use

Table 9.6 Risk: offsite versus onsite

Principles	Offsite	Onsite
Risk		
Cost	Overall higher cost potential, more predictable	Standard bidding process brings waste, cost is unpredictable
Handling	Transit damage potential, cumbersome large-scale unit install	Multiple trips, smaller pieces for easier per install handling
Public perception	Negative	NA
Innovative	Greater innovation possible	More difficult to achieve innovation complexity
Safety to labour force	Safe indoor labour conditions	Statistically more dangerous
Tolerances	Discrepancy between onsite and offsite elements present problems, element tighter tolerances	Tolerances can be accommodated easily in onsite installation
Fit	If not fit, changing size of elements costly	Onsite accommodation of fitting issues resolved without added cost
Quality	When increased, risk goes down	Higher exposure to risk due to material and joint failure

Based on Smith (2010). Used with permission.

your 'dream team' for early engagement and collaboration. Essentially, get the design team, construction team and manufacturing teams into a room on Day Zero to work out how the coordination of the project and the data assets will run.

The best application for prefabrication methods is where there is repetition such as in multi-residential, hotels, student accommodation, classrooms, aged care facilities, detention centres, etc. New build and renovation opportunities are available for prefab. Timber is well-positioned as a lightweight system for refurbishment and building extensions.

> With mass timber we can now reimagine real estate markets and build a city above a city. Urban and city densification includes construction air rights space, particularly in urban in-fill areas. As timber is lightweight (only 20% the weight of reinforced concrete) it allows for the construction of transfer decks over stations, rail yards, on top of car parks etc. providing a cheaper and easier way to construct building solutions with reductions in the size of vertical support structures and foundations.
>
> (Kremer, n.d.)

Figure 9.3 Assembly of panel/column system.
Source: Hayes/Marriage.

Figure 9.4 Assembly of full panel system.
Source: Hayes/Marriage.

Figure 9.5 Assembly of volumetric system.
Source: Hayes/Marriage.

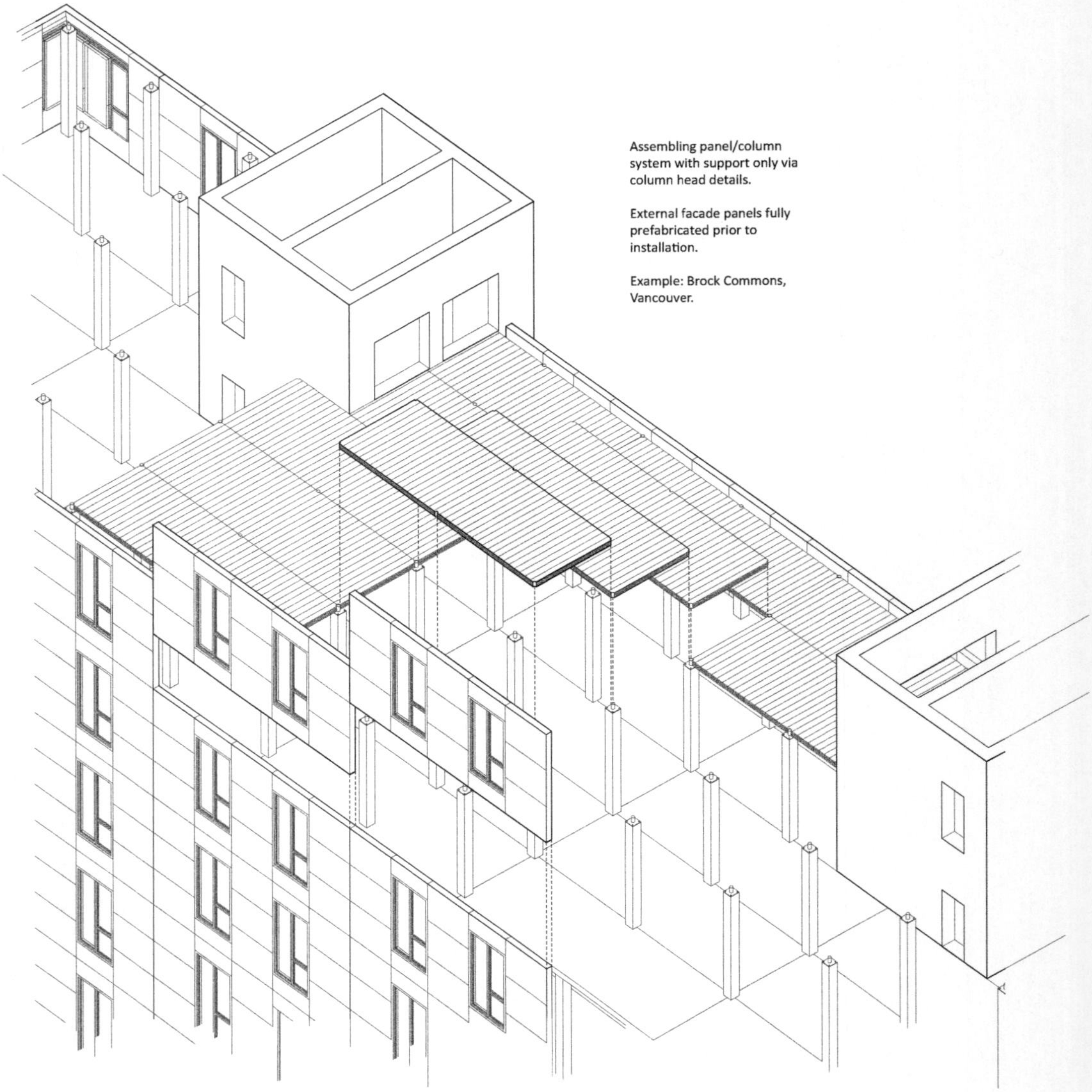

Figure 9.6 Floor slab and façade panel assembly.
Source: Hayes/Marriage.

Design, specification and manufacturing decisions need to be frozen in the design model, whether digitally in Building Information Management (BIM) systems or on paper outputs. Continual client changes will erode the benefits of prefab, as will any late-stage decisions to shift from traditional to prefabrication methods. There are online resources to assist with working out 'where and when' to use different prefab typologies. Use the PrefabNZ Material Matrix for key considerations to identify in early project stages (Figure 9.7).

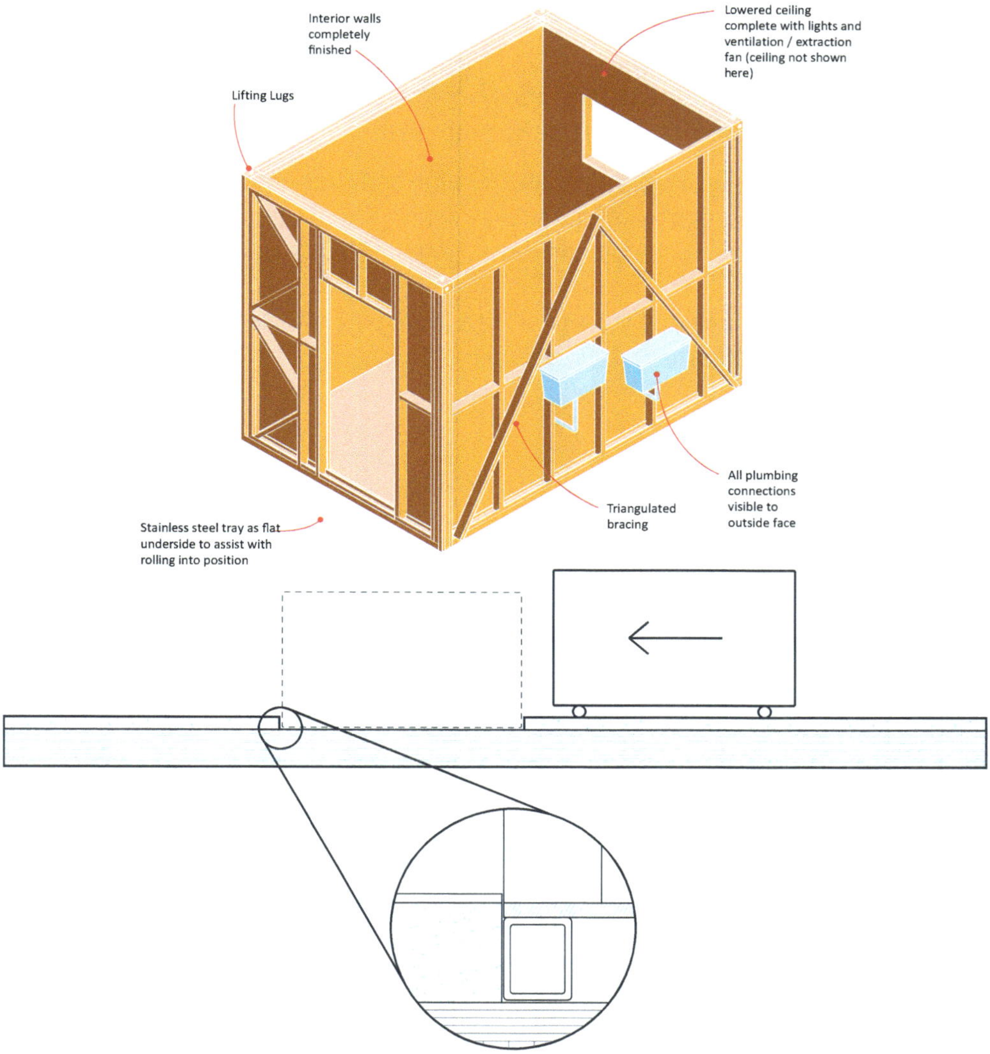

Figure 9.7 Prefabricated bathroom module.
Source: Hayes/Marriage.

Conclusion

This chapter provided an overview of terminology, clarification of key terms, and outline of how to evaluate when prefabrication is a strategy to explore over traditional construction

Five key areas to be aware of can be summarised as:

1. Assemble your 'dream team' early (perhaps we mentioned this already...).
2. Employ Lean methodologies from the beginning.

3. Establish the Shared Data Environment (SDE) for digital collaboration, early.
4. Use a Quality Assurance (QA) framework that interfaces with local building regulations and compliance processes.
5. Allow the team to 'fail fast' by testing, iterating and problem-solving together on the lessons learned. You may even want to set up the project with an in-house research and development (R&D) incubator.

Together these terms, tips and tables will be a great place to start on your prefabrication project journey – best of luck!

Works cited

Bell, Pamela. 2009. "Kiwi Prefab." Master of Architecture thesis. Victoria University of Wellington, NZ.

Bell, Pamela and Southcombe, Mark. 2010. *Kiwi Prefab.* New Zealand: Wellington.

CLT Handbook USA. *Think Wood.* Available at: https://info.thinkwood.com/clt-handbook

Kremer, Paul. *LinkedIn.* https://www.linkedin.com/pulse/designing-mass-timber-what-dont-teach-you-paul-kremer/

Ministry of Housing, Communities and Local Government (MHCLG) Joint Industry Working Group on MMC. 2019. *Developing a Definition Framework.* Working Group Members Include Buildoffsite, Homes England, National House Building Council (NHBC) and Royal Institution of Chartered Surveyors (RICS). Available at: https://www.gov.uk/government/publications/modern-methods-of-construction-working-group-developing-a-definition-framework

PrefabNZ, n.d. Available at: https://www.prefabnz.com/

Smith, Ryan. 2010. *Prefab Architecture: A Guide to Modular Design and Construction.* Hoboken: John Wiley & Sons.

10

Chapter Ten

Residential core design

Residential core design

Guy Marriage

DOI: 10.4324/9781003123873-10

Introduction

The design of any core is crucial to the good functioning of any tall building. In any building core there is a need for stairs, lifts (elevators), water pipes both in and out, electrical and data cables; but a commercial office block will also typically require the movement of a large amount of tempered air. Structural/services cores for residential buildings are inherently much simpler than those in commercial office buildings. In a residential apartment building, there is usually no need for large volumes of air movements, and generally far fewer services requirements overall, considerably reducing the volume required at the core.

Core design

Any building project is torn between two ideals: the need to have a tight, efficient plan for the sake of economy, versus the desire to have a building that creates enjoyable spaces for living. Generally, that means you don't want to waste space on the core design and you do want to limit the length of corridors as much as possible. Conversely you may want a common space which can be used by all residents of the building. The old European courtyard buildings are focused around an atrium/central space usable by all and achieve great sociability as a result. This is much harder to accomplish with a slim vertical tower, although even in the crushing density of the Hong Kong towers, common areas are still created to allow some room for community activity.

Despite a decade or two of *Friends* and *Seinfeld* depicting New York apartment living, people will often not want to be 'ambushed' near their actual apartment front door. The ground floor entry lobby is the most frequently-used common space for socialising, so allow some room for this social interaction to happen. While in New York this foyer also has room for a concierge, that is now unusual in world terms. Most apartment buildings these days have an unstaffed entry lobby: very few have a superintendent (Figure 10.1).

Stair design

Escape stairs (or fire egress stairs) are crucial to an apartment building. When we sleep our senses are reduced (no sense of smell and reduced sense of light and sound) and as residential buildings contain sleeping people, then the escape stairs are the most crucial part of the building's emergency plan. Lives truly depend upon these stairs.

Always provide at least two means of escape – this could mean two staircases intertwined in one core, or two separate cores with two separate stairs, but ideally the entry points to the two escape stairs should be far away from each other. The two staircases need to be available to everyone

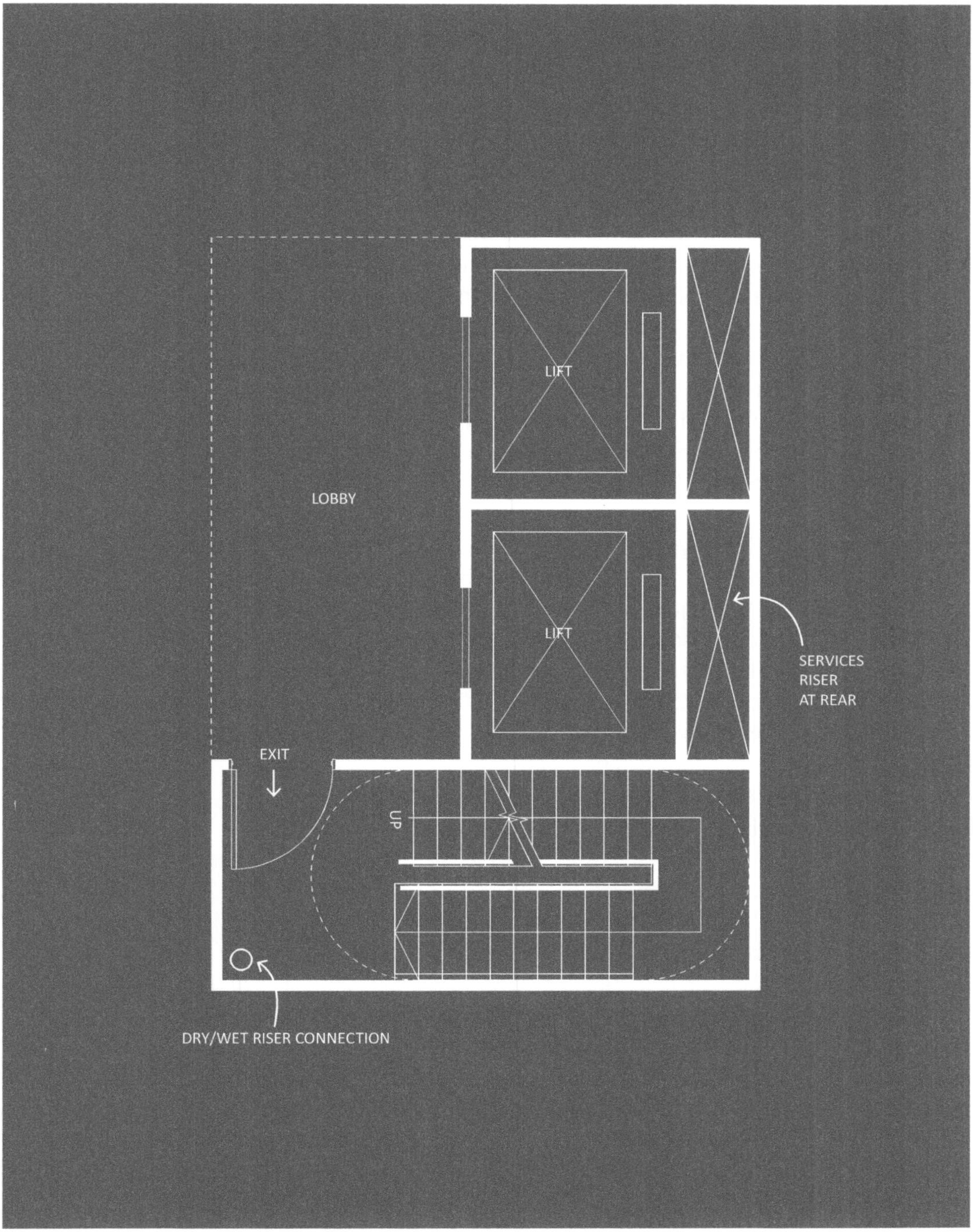

Figure 10.1 Core design.
Source: Hayes/Marriage.

in that building, so that in the case of an emergency (like a fire in the building) where potentially one stair could be blocked, there is always an alternative escape path. Some countries may allow you to design the building with only one stair but after the tragedy of Grenfell Tower in London, where the only (single) stair was inaccessible to many and 72 people died, then I don't think anyone wants that responsibility weighing on their minds.

These egress stairs are not just for escape though – they also function as an 'urban gym'. Many residents of high-rise buildings may prefer to take the stairs for a number of floors rather than use the lift/elevator. They will be highly used especially for the lower floors of an apartment building, so they need to be functional as well as comfortable spaces to be in. Encourage people to use the stair by making it a nice, safe, easily-accessible and enjoyable space to be; not a scary, remote back route with twists and turns. Concentrate on light, colour, acoustics, and materials, and provide daylight too if at all possible (Figure 10.2).

Stair widths are governed by the Building Code but as occupant loads in an apartment building are far lower than in an office building, the widths dictated may turn out to be quite narrow. A mandatory 900mm minimum (2′11″) width of a stair may sound reasonable but remember that the stair also needs to cater for occasional furniture moving and comfortable physical distancing so, aim for a minimum of 1,100mm (3′7″) at all times for apartment stairs, or even more if you can get it. The steepness of pitch is another factor – while an egress stair is permitted to be at a steeper pitch than a common access stair, it does not have to be – it can still be designed to comfortable accessible standards, if you want.

Pressurisation

Design a stair pressurisation shaft into the core if you can, where air can be pumped steadily into the stair itself, keeping it at a slightly higher positive pressure than the rest of the building. This will keep any smoke out of the stairwell and will save lives in the case of a fire. Stair pressurisation ducts are common on large office buildings but less common on apartment buildings, although this will change as architects come to realise how important a device this is for safety. The duct should connect to the stair shaft every couple of floors and be powered by a small fan at one end, pushing air into the stairwell. As doors are opened and closed under normal use, the pressurisation will also have the effect of helping keep the stair doors closed.

Scissor stairs

The concept of scissor stairs allows two stairs to run in overlapping spaces, with separate entries to the stairs from opposite sides of the core. The overall size of the scissor stair is longer than a single stair, but not as

Figure 10.2 Stair design.
Source: Hayes/Marriage.

large as having two completely separate stair shafts. This saves space in the building – and so also saves money for the developer. A scissor stair is only slightly more complicated to design than a single stair and if planned well, the half landing can be a structurally shared between the two stairs. The half landing needs structural support – at least a beam at the midpoint between floors (Figure 10.3).

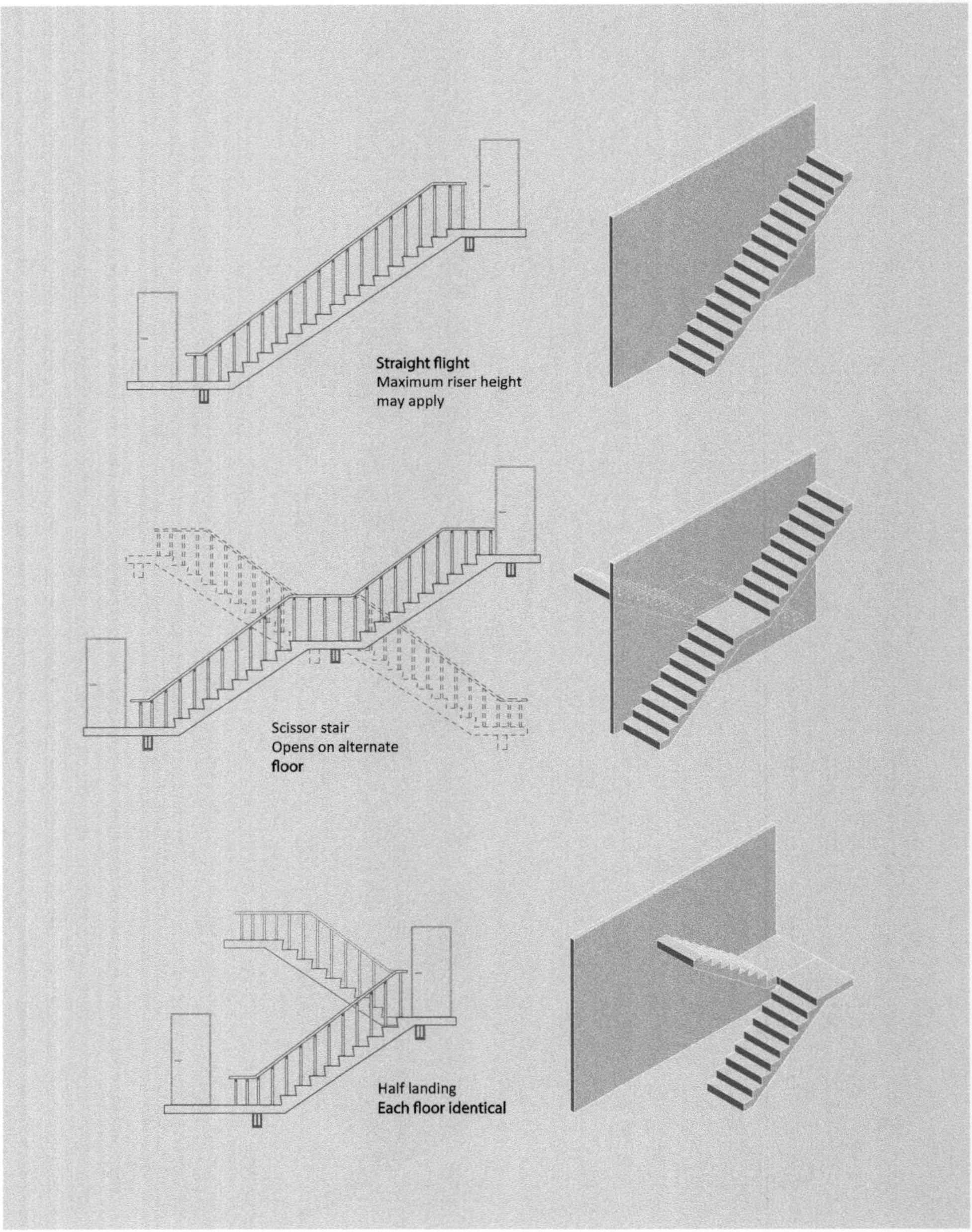

Figure 10.3 Stair sections.
Source: Hayes/Marriage.

Stair doors

No fire stair will be a safe place unless the stair doors are also fire-resistance-rated. The doors should always be shut, as they are fire doors. Make sure that they are not wedged open: again, this will save lives. Fire doors should have a similar Fire Resistance Rating (FRR) to the walls of the egress stair enclosure itself, i.e. if the walls are rated FRR 60/60/60 then the door will need to satisfy FRR-/60/60 (structural adequacy/integrity/insulation). Check this advice against your local Building Code as requirements differ between countries.

Depending on your Fire Plan, if you have a long corridor, this can be broken up with smoke stop doors. These can usually be held open with automatic closing mechanisms, held open with electro-magnetic latches that will release automatically in the case of fire alarm, or power failure. Please note: smoke stop doors are not fire doors.

Doors always need to open INTO the stair, in the direction of travel, except (obviously) on the ground floor where the exit doors should open OUT, i.e. again in the direction of travel. Lighting, acoustics and handrails both sides of the stair are also vital points to make sure that the space is comfortable and welcoming to use.

Other factors to design in are sprinkler positions, floor numbering and attention to ease of opening – like door handles. Nobody wants to touch those things again since Covid arrived. Choose handles that can be opened by an elbow or if possible, by a simple push plate. Choose RFID unlocking for everyday handling in place of manual touching of surfaces. Emergency use will always need to be manual (Figure 10.4).

Structural core integration

The structure of the core is often also the major place of structural support to the building. In an all-concrete building, the concrete walls of the core are used to gain sufficient lateral support and perform as shear walls. Structural cores are usually constructed first, leading the floors by weeks or months. In the case of concrete structural cores this is because in-situ concrete is laborious in time to prepare the reinforcing, as well as needing 28 days to come to full strength. Even so, modern slip-forming (or jump-forming) techniques will permit construction of one floor per week when the project team is running smoothly.

Do cores have to be concrete? In some countries, yes; in other countries, no. The response varies hugely depending on what jurisdiction your building sits under, as well as what the engineer and architect decide. In countries where a performance-based building code is in play, the core will need to comply with a fire-rating requirement, that can be met many

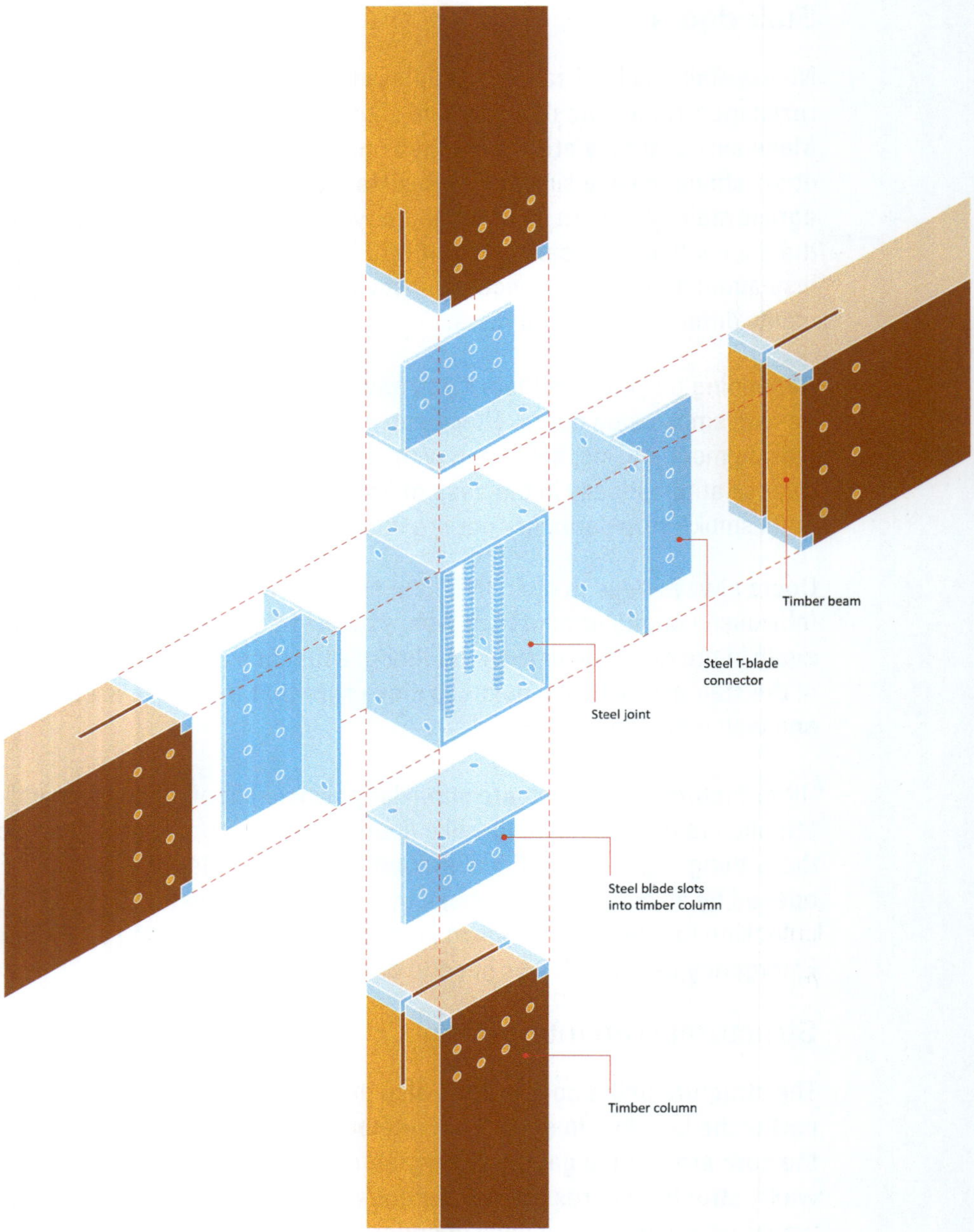

Figure 10.4 LVL beam to column detail with reinforced steel boot.
Source: Hayes/Marriage.

different ways: this may just mean a steel-framed structural core with FRR plasterboard wall framing to create the core walls. Alternatively, a concrete-walled core indicates that the core is part of the structural solution as well as the fire-resistant strategy.

A mass timber building may have timber core walls as well, but so far this has been kept mainly to low-rise buildings. Taller timber buildings often feature concrete cores that are the main structural resistance as well as providing fire resistance. Concrete cores are, after all, stiffer than timber core walls. Tallwood House at Brock Commons in UBC, Canada, prominently features two tall concrete cores that were constructed first, with the CLT flooring and PSL columns being constructed in a separate phase after the cores were complete. Here the concrete cores are fulfilling many different roles: structural stability, lateral stiffness, fire egress and refuge, as well as achieving compliance for the project with the Vancouver building authority (Figure 10.5). In New Zealand, recent work by Naylor Love has established that probably the most cost-effective solution can be gained by having timber take the gravity load, while steel structure can resist the lateral load (Naylor Love, 2021).

Lifts/elevators

You should always aim for a minimum of at least two lifts in the core, so that if one needs servicing then the other can still be used. Lifts in apartment buildings are used much less than in office buildings due to the far smaller occupant load and the limited times that people use them. Typically they will be at their busiest in the evening after work as people may be returning with shopping and groceries. A residential lift needs to be solidly built (as they are often used for furniture removal), reliable and convenient. It also needs to be right next to the stairs, so that residents can easily choose to use the stairs instead, if the lift is not coming. For any large scheme, designate one lift to function as a goods lift: furniture removal is the greatest threat to a lift's longevity. Careful design is crucial. The Covid virus has emphasized the importance of not touching shared surfaces like lift buttons, so expect a big push for voice-activated lift services in the future, as well as far better extract ventilation in the next generation of lifts.

Services core integration

The core is also the logical place to situate any of the main services that will rise up and down the building. Never run services from one apartment through the ceiling of the apartment below: this is illegal in many countries, but impractical and dangerous in all situations. Instead, route all services to the ceiling of the common area, or directly into vertical riser shafts that descend through the entire building. The ceiling in the common

Figure 10.5 Steel boot connector detail.
Source: Hayes/Marriage.

areas may often need to be a little lower than the apartments themselves, in order for services to slip comfortably beneath the layer of structure connecting back to the core.

Although there is little requirement for common air-handling services in a residential building, there are still many services required, mostly in the form of pipes and wiring. These include the following.

Power and data

- Electrical wiring – for power to each apartment. This will end in an electrical distribution board (DB) in a cupboard at each floor. This cupboard may also contain:
- Emergency lighting. A separate wiring network will supply the emergency lighting. Alternatively, the emergency lighting may be individually battery-powered in case of mains power outage.
- Fire alarm cabling – another separate set of wiring, usually with red cables. Connects to the fire alarm sensors and to the fire warning sound system – which may be alarm bells or a recorded message.
- Security cabling. There may be a separate small network for security cabling, connecting alarm sensors.
- Data wiring – will often also end in a separate small cupboard at each floor. Often, competing broadband networks may connect their fibre at this point. Generally, keep the power riser separated from the data riser. Seal off all cables penetrating through from one fire compartment to another, with fire-rated sealant or fire collars, without exceptions.
- Cables and pipes will run within the ceiling cavity in the common areas, until they reach the apartment wall and enter into the private residence within.
- Each apartment will then have their own fuse box or switchboard (SWB) within the confines of their apartment.
- Risers containing electricity should always be separated from those containing liquids, for obvious reasons. The power and data wiring can be threaded through holes cored in each floor slab and must be fire sealed at each floor with fire collars.

Fire suppression

- Sprinkler supply – sprinklers work by having the piping always full of water, therefore it has to remain primed, ready to go off. This is on a completely separate circuit to any other water supply. Any tall apartment building should logically be fitted with sprinklers for human life safety reasons: all new buildings will have them.

- Dry riser/wet riser – pipes for the Fire Service to connect to in the case of a fire, with riser outlets on each floor in the fire stairs. The pipes can be either 'dry' risers or 'wet' risers, depending on whether or not they are pre-charged with water. Local fire regulations may dictate whether your building has a dry or a wet riser.
- Older buildings may typically have an emergency hose reel on each floor but modern fire-fighting techniques prefer the occupants to summon the fire service quickly. There needs to be at least one fire emergency point per floor connected direct to the local fire service. Powder fire-extinguishers are still permitted for limited fire suppression.

Piped liquids

- Fresh cold water – drinking quality water supply to kitchens and bathrooms.
- Heating pipes – these will circulate hot water from a central boiler in the basement, to warm each apartment via radiators. Common in northern Europe and other cold climates.
- Projects with Community Heat and Power (CHP) may get hot water pumped directly into the building from neighbourhood excess energy.

Plumbing and drainage

- Grey water – slightly dirty used water from hand-basins and showers, can only be used for supplying the toilet cistern. Unusual in apartment buildings, as it would require grey water storage in each apartment; however, with the predicted growth in population across the world and the ongoing over-consumption of fresh potable water, this is something that architects and hydraulic engineers will have to plan for.
- Black water – the sewer pipe or soil stack going down and out, connecting at ground level to the public sewer. Composting toilets are unlikely to be specified in apartment buildings.

Other services

- Gas piping – gas cookers are often requested in apartment buildings but need to be installed very carefully and in many cities, they are banned completely. If someone leaves a gas cooker on and it combusts, the end result to the building could be catastrophic. The Ronan Point disaster is the classic example of why gas is not good in an apartment building (Heathcote, 2020). Generally, any gas pipe will need to be installed in a well-ventilated riser so that any gas leak cannot build up. As an alternative, a charcoal barbeque on a balcony may be a good option for those who want their meat seared with a smoky flavour. Maybe this is just a

New Zealand thing, but our apartment building sizzles with the sound of multiple barbeques being fired up on warm summer evenings!

Air handling

- Air ducting may be needed to some areas even if the building is naturally ventilated. Air supply systems may be required to basement car-parking, to stair pressurisation, and especially to any possible mixed-use lower floor purposes such as ground floor retail. Due to an apartment building being divided into many different fire compartments, then any common air distribution is inherently risky and not recommended.
- Fresh air supply – from outside, to each apartment. This may be centrally supplied as noted above, but in most cases fresh air is locally/individually sourced to each unit, and the intake is carefully placed to avoid any mixing with the exhaust air.
- Exhaust air extract – in an apartment building, extracted air is always expelled at each apartment direct to the external air from three key areas: bathrooms, laundries and kitchens. Generally, as long as air is being exhausted to the external environment, replacement air will make its way in to replenish the air expelled although if the building is particularly well sealed, then an open window or relief ducts may be necessary.
- Staircase pressurisation duct (SPD) – ensures air pressure in the stairwell is always positive to stop any smoke infiltration, as people need to evacuate safely through the stair. One SPD for every staircase.

Miscellaneous core activities

- Door bells – the importance of the door bell is waning but not over yet. A video intercom at the front door is essential for friends to announce their presence, but mobile phones are providing a more common way of interacting now.
- Deliveries – often parcels will not be delivered to your door but just to the front entry lobby, and parcel couriers need both a place to deliver the goods and to wait for your arrival. Food deliveries, of course, typically make it right to your door, while the post does not.
- Goods lift – if the apartment complex is large enough, design in a special large, rugged lift just for furniture removals etc. This both keeps the regular lifts available for the general population while also stopping the lift get broken from overuse. Make sure it has easy access to the freight loading zone outside where a truck can safely park.

Rubbish

There is no agreement on rubbish chutes. Some cities mandate their use, while others do not encourage them. The problem with rubbish chutes is that they get abused and they require constant maintenance to stop them smelling bad. While in theory they effortlessly transport rubbish down to a patiently-waiting rubbish bin many floors below, in reality they get dirty, they get blocked, they are a source of noise and they are a route for fire to spread. Check with your local City rules / Building Code first.

Linked closely with the possibility of a rubbish chute is the provision for storage and collection of rubbish and recycling. That space needs to be clearly accessible by the collection trucks, of which there are often several different companies. One collects bottles, while the other cardboard and paper; another collects plastics for recycling, while yet another (more frequently) collects the trash. This typically involves large trucks that lift the bins above their heavy snorting bodies – the ground floor collection space needs to be outside (or nearly so) and enclosed for acoustic reasons if at all possible. The noise and smell from the rubbish trucks may well devalue the price of apartments near the bins, so design very carefully with that in mind.

Amenities

If your apartment building is large enough, plan for additional amenities on site such as shops, groceries, gymnasium, swimming pool, barber, hairdresser, restaurant or coffee shop. The ground floor is the best place for these facilities, so that they can also provide for passing public foot traffic, although in some mega-apartment buildings then additional shops may be on other levels with internal sky-gardens or atria. Ground floor shopping should be flexibly designed to appeal to any retailer, although ideally leased to a function that the building occupants will need.

Works cited

Heathcote, Christopher. 2020. "When modernist Britain fell down." In *Quadrant,* 82. Sydney: Quadrant.org.au.

Naylor Love. 2021. *Building Carbon Calculator.* Available at: https://www.naylorlove.co.nz/carbon/

11

Chapter Eleven

Apartment planning

Apartment planning

Guy Marriage

DOI: 10.4324/9781003123873-11

Introduction

There are many – maybe endless – ways to lay out the basic rooms of a house. They will vary from suburb to suburb, from country to country, and from one climate condition to another. House plans may vary considerably between different cultural groups in different countries. However, in multi-unit building projects that wide range of choices get narrowed down to just a few prime variations. Unlike houses which have been designed individually for a person or a family, apartment units are designed and built *en masse*, usually for a developer or a local government authority, and therefore usually without ever knowing the actual final inhabitants. The end result is that apartment buildings are remarkably similar the world over – their common feature is that they are shaped by money. Hopefully, more than just money, but hard-nosed commercial aspects rule the equation here.

Extensive market research is done first by the architect and the design team, to establish the desired mix of apartment sizes, numbers of bedrooms, and cost targets. Sometimes this will result in a project that is straightforward: X one-bedroom units, Y two-bedroom units, and Z three-bedroom units. Avoid solutions that propose all of one type of unit, except perhaps if it is a student hostel – and even then, a range of bed numbers is better than a monotonous solution all the same. A range of unit types will ensure a range of people living in the building, which makes for a better urban community: young, old, rich, poor, brown, white, with children or without. That mix of residents makes for a good urban village and brings the building to life.

Apartment sizes and indicative costs

Earlier chapters explored basic rules of block planning and site orientation. With good site planning, your apartments will have already been laid out in a sensible manner, allowing all apartments some much needed sunshine and ample daylight. At this stage of planning, a broad outline target is established to the possible costs: developers will be looking at a blanket rate of X dollars/pounds/euros per m^2. Remember that the sales cost to an eventual buyer relies on a lot more than just the construction cost – the developer will have paid a hefty amount for the land, will have to pay development fees to the Council as well as extensive fees to the design team; all before they even engage the Contractor. Often this formula works in reverse: the developer will know what market price will be achievable for a finished apartment and so the price is reverse engineered. Sometimes, the design team fees get cut to a minimum, sometimes it is the appliances budget; mostly, it is the landscaping.

World house sizes

As noted in Tables 11.1 and 11.2, market expectations of home sizes vary considerably from country to country (Marriage, 2010). Some countries have spatial controls to set a minimum size, while other countries still believe that letting the free market rule is best. European housing standards are generally more generous than those in the UK, which were for many years informally guided by the 'Parker Morris' standards (Morris, 1961).

Countries that formerly relied on large suburban section sizes are now rebuilding large swathes of their suburban tracts: the mythical 'quarter-acre' section (1,012m^2) of New Zealand has long had its day, as section sizes shrink to a third of their former size, or less (Marriage, 2010). Paradoxically, for many years while average house sizes were growing larger, section sizes and side/front/back yards were growing ever smaller (Marriage, 2014). While builders and developers have found more profit over recent years in building ever larger houses, the upcoming boom in apartment building will see average sizes become more realistic. Apartments need to be usable by singles, couples, and increasingly, whole families. As such, a range of sizes needs to be provided.

Table 11.1 Minimum apartment size versus comfortable apartment size

	Studio (1–2 persons)	**One bedroom (1–2 persons)**	**Two bedroom (2–4 persons)**	**Three bedroom (3–6 persons)**
Minimum area in m^2 (sq ft): Hong Kong	19 m^2 (205 sq ft)	40 m^2 (430 sq ft)	50 m^2 (538 sq ft)	55–60 m^2 (592–646 sq ft)
Minimum area in m^2 (sq ft): Great Britain	–	39 m^2 (420 sq ft)	61 m^2 (657 sq ft)	74 m^2 (797 sq ft)
Recommended minimum area in m^2 (sq ft): New Zealand	40 m^2 (430 sq ft)	50 m^2 (538 sq ft)	75 m^2 (807 sq ft)	100 m^2 (1,076 sq ft)
Comfortable size in m^2 (sq ft):	n/a	60 m^2 (646 sq ft)	90 m^2 (969 sq ft)	120 m^2 (1,292 sq ft)
Luxurious size in m^2 (sq ft):	n/a	100 m^2 (1,076 sq ft)	160 m^2 (1,722 sq ft)	220 m^2 (2,368 sq ft)

Sources: Auckland Council, British Standards, DCLG, and Marriage.

Table 11.2 Average house size, global survey

Country	Average (all houses) in m^2 (sq ft)	Average (new houses) in m^2 (sq ft)	Average (new, detached) in m^2 (sq ft)
Australia		206 m^2 (2,236 sq ft)	235 m^2 (2,554 sq ft)
Austria	91 m^2 (989 sq ft)	96 m^2 (1,043 sq ft)	
Belgium	86 m^2 (935 sq ft)	119 m^2 (1,293 sq ft)	
Canada			177 m^2 (1,919 sq ft)
Denmark	109 m^2 (1,185 sq ft)	137 m^2 (1,489 sq ft)	
Finland	77 m^2 (837 sq ft)	81 m^2 (880 sq ft)	
France	88 m^2 (957 sq ft)	113 m^2 (1,228 sq ft)	
Germany	87 m^2 (946 sq ft)	109 m^2 (1,185 sq ft)	
Greece	80 m^2 (870 sq ft)	126 m^2 (1,370 sq ft)	
Ireland	88 m^2 (957 sq ft)	88 m^2 (953 sq ft)	
Italy	90 m^2 (978 sq ft)	82 m^2 (891 sq ft)	
Japan	95 m^2 (1,033 sq ft)	132 m^2 (1,435 sq ft)	187 m^2 (2,033 sq ft)
Luxembourg	125 m^2 (1,359 sq ft)	101 m^2 (1,098 sq ft)	
Netherlands	98 m^2 (1,065 sq ft)	116 m^2 (1,261 sq ft)	
New Zealand		176 m^2 (1,913 sq ft)	200 m^2 (2,174 sq ft)
Portugal	83 m^2 (902 sq ft)	82 m^2 (891 sq ft)	
Spain	85 m^2 (924 sq ft)	97 m^2 (1,054 sq ft)	
Sweden	90 m^2 (978 sq ft)	83 m^2 (902 sq ft)	
UK	85 m^2 (924 sq ft)	76 m^2 (826 sq ft)	
USA		203 m^2 (2,204 sq ft)	221 m^2 (2,407 sq ft)

Sources: Demographia, Japan Statistical Yearbook, European Housing 2002, Australian Bureau of Statistics, Canadian Home Builders Association, Infometrics.

Floor spans

Residential apartment building structures are completely different from office buildings. Office construction tends to concentrate on maximising wide spans to gain large flexible areas of floor space, but in apartments the floor planes are deliberately broken up into much smaller spaces to form self-contained spaces, meaning that structural columns can be much closer together as well. Apartment walls do not need to be flexible in their position – once built, the apartment wall is fixed rigidly in position for the life of the building itself. Strata titles/unit titles require a fixed, permanent position in space to create legally binding property boundaries.

As a result, while an office building may be just a regular grid of columns, a core and an external façade, the construction of a residential building is highly compartmentalised by comparison. The conversion of an old office building or industrial warehouse is therefore seldom well-suited to use as a residential complex. The majority of the upcoming wave of apartments will need to be purpose-built, not converted from old C grade offices, despite the temptation to utilise unused office buildings abandoned during the Covid pandemic (Wall, 2020).

An office building will typically be designed to maximise open span floor spaces, with column lines set 9–12m apart (roughly 30′–40′). An apartment building is very different, with structural centres closer together, with more usually 4.5–6.0m (roughly 15′–20′) structural support spacing. In many cases, apartment widths are similar to more traditional two to three storey terrace housing, where overall widths are based on module spacing of 4.35m wide, 4.65m, 5.35m, 6.35m, or 8.35m (Levitt and McCafferty, 2019). These spacings have been in practice for centuries, carefully apportioning living space whilst also relating to the spanning capabilities of timber joists. But for apartment design, a module of 7.2m centre to centre is best: allowing for a bedroom and a living room to exist comfortably side by side.

Floor planning/spacing/structural integration

With the narrow residential spacing modules and modern means of construction, floor slabs in apartment buildings will be solid, not timber joists. The slabs do not have to span far, so the beams that support the slabs may be less deep, or in many cases structural floor beams are designed out altogether, so concrete floor slabs are laid simply as flat slabs and hence the floor to floor spacing can be much tighter. Note that while CLT (Mayo, 2015) and concrete can be laid as simple flat slabs, steel-framed buildings will almost always require beams – likely one beam every 3.0m or so for a steel/concrete composite floor.

Living space

The living room is the most crucial part of any apartment. It needs space, daylight, fresh air, and hopefully a decent outlook. Design in full height doors to enhance a sense of space and use full height mirrors to make spaces feel larger and to bounce light around the room. Design in solid walls as well as glazed frontage, to provide space to hang art and provide some visual privacy. The living room is often directly off the corridor or entry lobby to the apartment, so in many cases it can be a highly public space. Usually the space is combined with the dining and kitchen spaces as well – it is the centrepiece of the entire apartment, often with the other rooms leading off (Moussavi, 2014).

How big should it be? While the UK stuck firmly to spatial standards from 1967 to 1980, set by the Parker Morris report (Morris, 1961), in more recent years, the London Housing Design Guide has set new minimum requirements (Mayor of London, 2015), mirroring those of the 'nationally described space standard' (DCLG, 2015). The London Housing Design Guide recommends that the kitchen/dining/living area should be a minimum of 3.2m (10′ 6″) wide at the narrowest and to provide a total of 23m^2 for one or two persons, growing to 31m^2 for a house for six people. By contrast, the Auckland Design Manual indicates a kitchen/dining/living room space for an apartment should be a rectangular 3.8m $\times$ 8.0m (30.4m^2) or a similar-sized area within an L shape. That allows 4m (13′) as the living area and another 4m as kitchen with a slim dining table. The Good Solutions Guide for Apartments recommends a maximum of 8m depth for any single-aspect apartments, with a minimum ceiling height of 2.7m in habitable rooms (Sills van Bohemen, 2005). Some European spatial standards are clearly higher than Britain's: Denmark and Luxembourg seem particularly generous.

Living area projections: push out, pull back

Flat façades to apartment buildings often create residences that are unpleasant to live in. They're relatively cheap to build – but they can look cheap too. Adding depth and complexity to the façade can also create vibrancy and provide pleasure to the inhabitants within. After all, apartment living is all about being in the centre of urban activity and thus being able to see the life in the city streets below. If you want a more interesting building (both to look at and to live in), modulate the façade by creating projections and recesses. Push the living area out slightly in one area, pull it back in another, to create space for a semi-recessed balcony or deck. This will also give your living room the chance to create natural ventilation flow through the apartment.

Design living areas that can spread naturally onto an external deck and allow the residents to have variety in their possible living conditions. Sliding or folding doors onto the deck will allow a blurring of the line between inside and outside. This is crucial. Some people will live with the living room doors closed all the time and not use their deck, while other people (like myself) will leave a door or window onto the deck open all year round, to establish an easy transition from inside to outside. With the deck set back behind the edge of the projecting living room, this also allows the deck to open onto two different rooms and create ventilation for both (Sills van Bohemen Architects, 2005).

While a third of our lives may be at work and a third asleep, the remaining third is leisure time. Unlike in a suburban house where dwellers can have a choice between the living room and their garden, for apartment dwellers, the city itself is their living room. They may spend much of their free time in nearby cafes and restaurants. Apartment design needs to be tightly focused and not waste space so design out any corridor within the apartments. Every space should have at least two or three functions – flexibility is the key to success.

Kitchens

The kitchen is an area of the apartment that eats up money. I've put in kitchens that range in price from $5,000 each up to $90,000 – the cost difference is in the size and specification level (hint: granite and marble worktops can add significantly to the price). Kitchen layouts are typically: the galley (a single straight bench), the return (an L-shape bench), the three-sided (or U shape), the island bench, or combinations of those. There are many books and specialists on kitchen design, but for apartment building, my recommendation would be to have a kitchen plan that is simple and repeatable *en masse*, like a simple straight or L-shape bench with an island that can act as a gathering place or breakfast bar (Figure 11.1).

Situate the fridge and pantry at each end of the kitchen bench, with cooking hob and sink in between. Some people like to have the hob on the kitchen island – or a kitchen sink on the kitchen island – but these both really need to be on the same length of bench. The sink needs to be near the cooking top so that cooking utensils can simply go straight into the sink without leaving a trail of cooking debris across the floor. Leave the bench for chopping vegetables!

Ensure that the kitchen is near but not directly against the inter-tenancy wall, to simplify services like water, foul water, electricity, and extract noise. Inter-tenancy walls are usually fire-rated walls, so any penetrations are difficult and often forbidden, especially if the wall is concrete or brick or block – or even CLT. No matter what your inter-tenancy wall is made from, getting wiring and water-pipes into it or out of it is a nightmare, so build a thin false wall in front of it that you can happily penetrate with services. You can thank me later, when you are fitting out your kitchen on site.

Lastly, think about this: do you really even need a kitchen? For urban dwellers in the heart of the city, many will simply eat out most nights at a nearby café, restaurant, deli, or trattoria. My friend Tom has taken it to an extreme with his funky inner-city apartment: as he likes to eat out a lot he has turned his oven ('never used it!') into a wine-cellar instead. The simple answer is: yes, of course, you need a kitchen but it does not have to be a big expensive affair. Some of the pricier apartments for the obscenely wealthy have vast 'show' kitchens, complete with separate rooms for the real kitchen work; what used to be called a scullery. By contrast, in Italy they sell a 1.2m wide all-in-one kitchen that includes fridge, sink, cooking hob, microwave, and a cupboard; almost an item of furniture that you can take with you when you move. Italians know how to live a truly urban life!

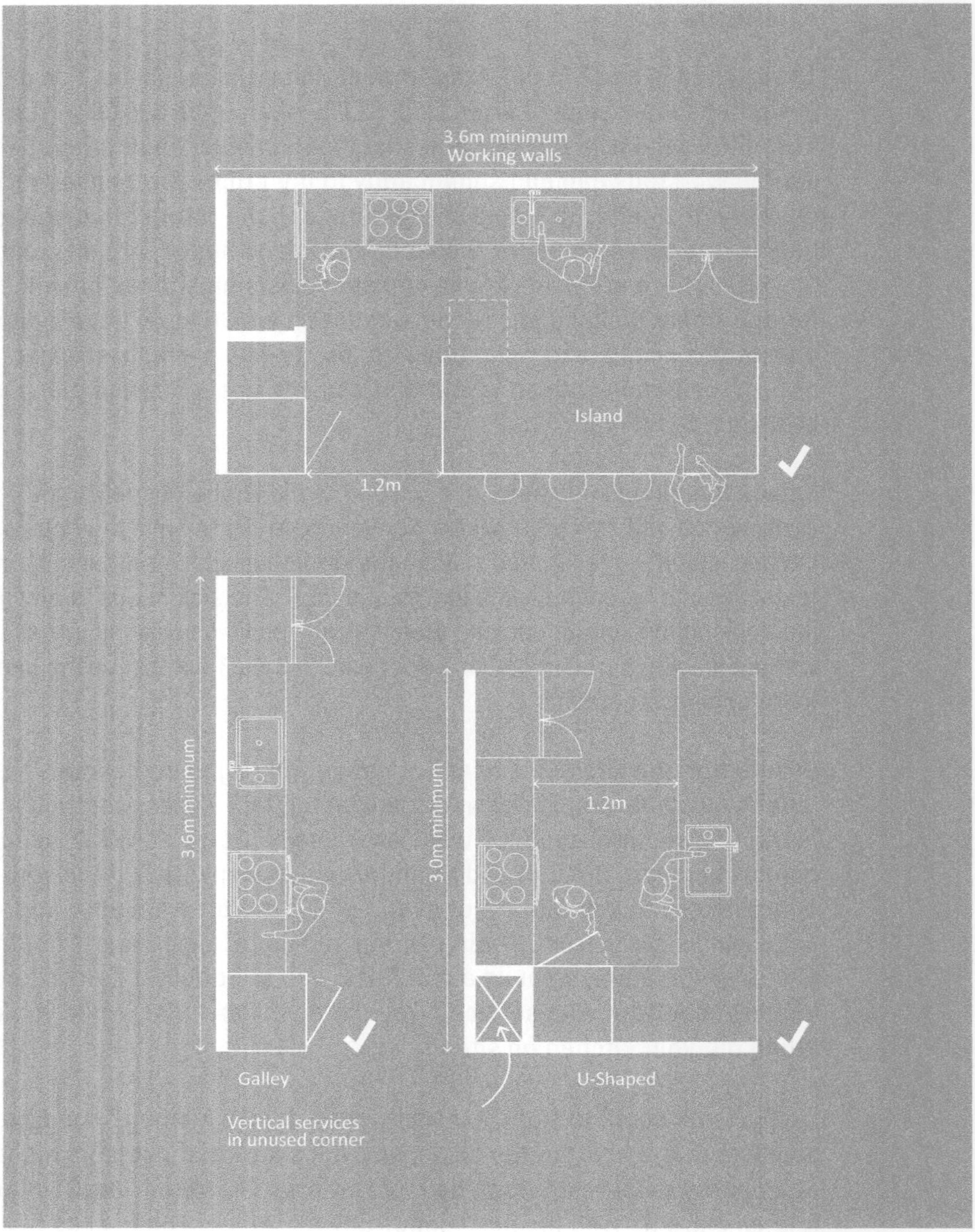

Figure 11.1 Kitchen planning.
Source: Hayes/Marriage.

Bathrooms

The most expensive part of any build (per m^2) is usually the bathroom – with more expensive finishes (waterproofing, tiling, grouting), more expensive services (hot and cold water, electricity, ventilation, drainage) as well as the bathroom fixtures themselves. Although still a luxury in many countries, having more than one bathroom is now fairly standard within apartments. Most units will now have two bathrooms for two to three bedrooms – perhaps one as an ensuite for the main bedroom and the other bathroom available to all household users. It is therefore important to get the bathroom design right, with maximum space for the people using the bathroom and minimal wasted space to help keep the cost down. Bathrooms are also the room most suitable for prefabrication as a modular volume. Get it right for one bathroom module and you've then got it right for all of them (Figure 11.2).

The planning and positioning of the bathroom services sets out where vertical plumbing risers will need to run through the entire building, so pay early attention to the plumbing and drainage due to the considerable effect they may have on your floor plan. Toilet fittings can come in a P-trap (which voids the sewer pipe out to the rear) or an S-trap, which requires the sewer pipe to go straight down through the floor. As you will want to avoid any pipes going into the apartment below, any S-trap toilet pans need to be installed over a false floor. P-trap fittings can be fitted against a wall so that the connection to the services riser can proceed, but make sure that is not an inter-tenancy wall.

There are a few simple classic designs for apartment bathrooms. For plumbing simplicity, arranging all sanitary fittings in a single row makes a lot of sense. The plumbing in the wall can supply all fittings in a short run, and waste water can be piped away simply. It can also work well as a disabled-friendly design – Universal Design – creating equal access and availability for users of all sizes and abilities. The size of a bathroom like this is usually about 2.7m (9′) long and about 1.6m (5′) wide. Universal Design requires a clear 1.5m diameter circle to allow wheelchairs to turn around.

Where space is at a premium a simple square bathroom may appeal, with fixtures clustered on two sides. The plumbing in the walls has to go further and will probably need to run under the floor, so designing in a raised area of flooring can help. A simple 1.8m × 1.8m plan can work well, but results in a small shower. Stretching it to 1.8m × 2.4m can accommodate a small bath instead. Some units are even smaller – down to 1.5m × 1.5m – but for most people, that's barely big enough for a WC, basin and shower, let alone enough room to turn around in. I would suggest that when you're naked and vulnerable in the bathroom, the very last thing you would want is to be banging into the walls and door handles.

There is also a skinny alternative – the long, thin, double-ended bathroom if you are really stuck for space. This simple layout puts a shower at one end, a toilet at the other end, and a wash-basin in the middle between them. The door opens from the centre. They can be installed in a space 900m wide, but I suggest a minimum of 1,200m wide instead unless you

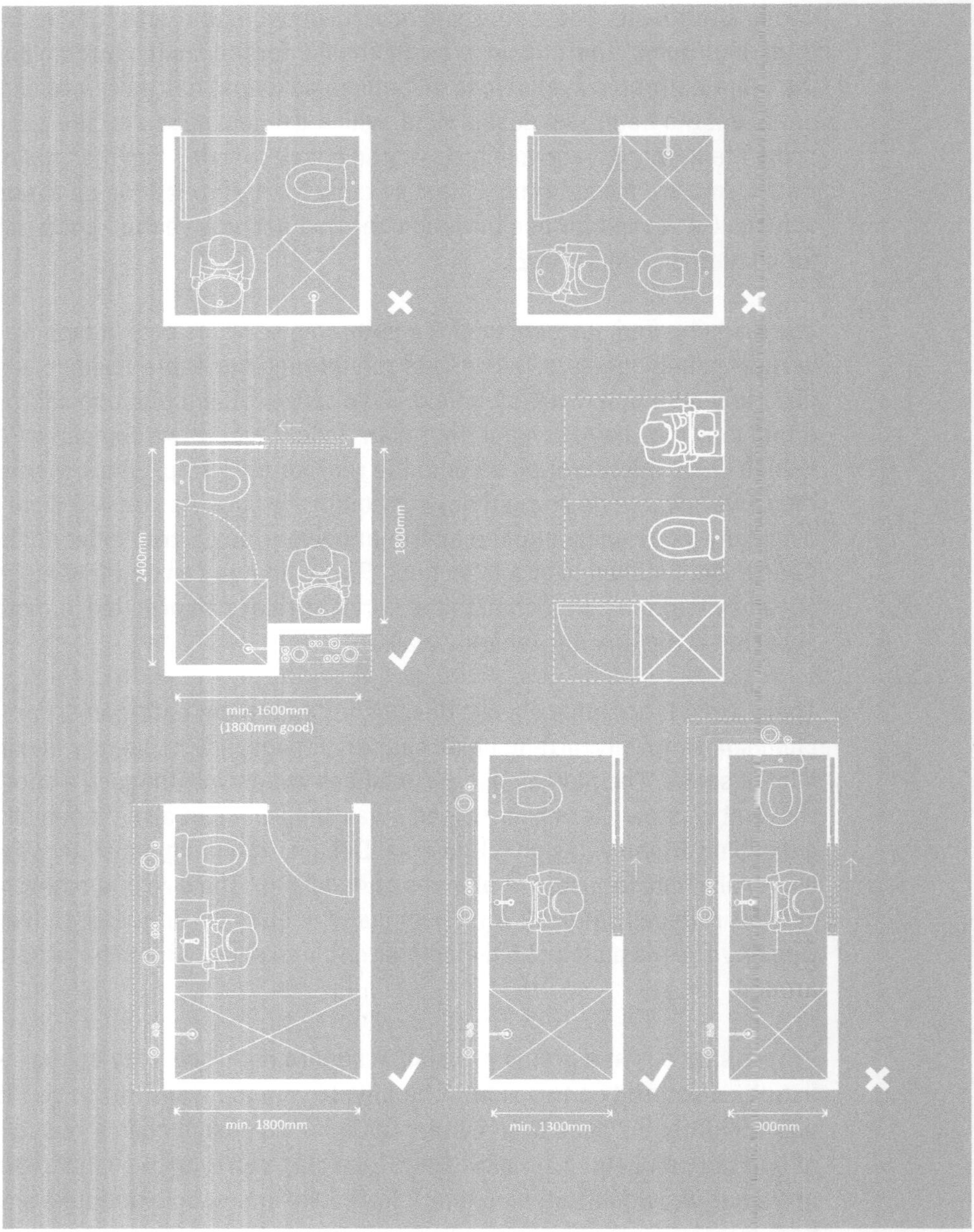

Figure 11.2 Bathroom planning.
Source: Hayes/Marriage.

want parts of you sticking out the bathroom door when you clean your teeth. Typical room length of these is about 3m. Note: if you are planning to put the laundry tub/washing machine in any bathroom, then the room will need to be bigger. There's rarely enough room to have a completely separate laundry in an apartment – make use of a cupboard instead, but make sure it has an adequate overflow drain! Always provide floor drains so the bathroom cannot flood into the apartment below. In some parts of the globe, for some odd quirk of history, washing machines are placed in the kitchen. Yuk. In my world, kitchens are for fresh food, not dirty laundry.

Bedrooms

Let's face it: the bedroom is where we spend about one third of our daily life, but when we are mostly asleep. Bedrooms are tricky – they need to be situated by an openable window to ensure the possibility of fresh air, but they also need to be somewhere quiet to allow the possibility of deep sleep. Darkness too is important, and not always easy to achieve in an inner-city apartment with all those blinking city lights. Curtains, blinds, or shutters will be vital, as will adequate wardrobe space and storage options.

If you can, allow the bedrooms to open out onto a private deck or balcony. Allow room for dressing and undressing, most of all, allow room for privacy. This is not a hotel room: this is someone's daily life. Build in wardrobes are a must, to lessen the need to buy furniture. A classic response to adjoining bedrooms is the use of wardrobes or an ensuite to space them apart – acoustic separation is a good thing. Allow for enough wall space to hang art or to decorate; a completely glazed wall is not usually a good idea in a bedroom. Allow room inside for seating, somewhere other than the bed. If you have to resort to shrinking the size of the furniture to make it look spacious (it has been done too many times to mention), then you shouldn't be in this business.

Although it is not an official standard, I like to test bedroom sizes by circular centripetal force with a domestic feline, answering the age-old question 'is there enough space to swing a cat?' (Marriage, 2003). That is a circle with a diameter of 2.6m minimum (8′6″) in plan – your imaginary cat should not hit the walls. DCLG advise that their minimum room dimension is just slightly bigger, at 2.75m (DCLG, 2015). Another way of testing room sizes is by imagining living there yourself. If you couldn't stand the thought of living there long-term, then why expect someone else to have to put up with it? As architects, we have a moral conscience to design to a standard of quality, not just dance to the developer's tune.

Home office/study

In this post-Covid era, the need for an extra space to use as a home office is likely to change future apartment design. In fact, couples may now require two home office areas, preferably at opposite ends of the apartment for acoustic reasons. I can't stress enough that workers need to not be sitting in an office in the bedroom – a bedroom should be designed for sleeping in, not for working in or running a business (Kenway and Holden,

2020). For this reason, a simple alcove off the side of a circulation space can work well as a study and home office.

Storage

You may not think this, but well-designed storage is one of the most important factors in apartment design. Unlike a standalone house with a backyard, apartments generally do not have extra space that can be easily shut off and locked away. People tend to accumulate things, not all of which can be Marie Kondo'd away after use. Winter overcoats, summer dresses, work boots, skis or snowboards, fishing gear, even your luggage for the weekend away: everything needs a simple and accessible place to be stored. That includes things as bulky and yet everyday as the vacuum cleaner (awkward shape and heavy) as well as the spare duvet or a child's bassinet. If your apartment floor to floor dimension is tall enough, consider designing in some cupboards up high above the bedroom or bathroom ceiling – wherever you can. Even better, if you have the space, design in storage under a raised floor area, to give a large space that also doubles as sound proofing.

Works cited

Auckland Council. 2018. *Auckland Design Manual.* Auckland: Auckland Council.

British Standards: minimum gross internal floor areas sizes. 2015. *Technical Housing Standards – Nationally Described Space Standard.* London: Department for Communities and Local Government.

Department for Communities and Local Government. 2015. *Technical housing standards – nationally described space standard.* London: Department for Communities and Local Government.

Green, Michael. 2018. The Case for Tall Wood Buildings (2nd Edition). Seattle, WA: Michael Green Architecture.

Green, Michael and Taggart, Jim. 2017. *Tall Wood Buildings – Design, Construction and Performance.* Basel: Birkhauser.

Kenway, Peter and Holden, Josh. 11 April 2020. *Accounting for the Variation in the Confirmed Covid-19 Caseload across England: An Analysis of the Role of Multi-generation Households, London and Time.* New Policy Institute (NPI). Available at: https://www.npi.org.uk/files/2115/8661/6941/20-04-11_Accounting_for_the_variation_in_Covid_cases_across_England.pdf

Levitt, David and McCafferty, Jo. 2019. *The House Design Handbook: A Guide to Good Practice* (2nd Edition). New York: Routledge.

Marriage, Guy. May/June 2003. "No swinging cats in inner-city flats." In Architecture *New Zealand*. Auckland: AGM Publishing.

Marriage, Guy. 2010. "Minimum vs maximum: size and the New Zealand house." *In 2010 Australasian Housing Researchers' Conference*. Auckland: University of Auckland.

Marriage, Guy. 2014. "Pavlova paradise and the curse of the side yard." *In Building a Better New Zealand* Conference. Auckland: MBIE.

Mayo, Joseph. 2015. *Solid Wood – Case Studies in Mass Timber Architecture, Technology and Design*. London: Routledge.

Morris, Parker. 1961. *Homes for Today and Tomorrow*. London: Her Majesty's Stationery Office.

Moussavi, Farshid. 2014. *The Function of Style*. New York: Actar/Harvard University Graduate School of Design/FunctionLab.

Sills van Bohemen Architects. 2005. *Good Solutions Guide for Apartment Buildings*. Auckland: North Shore Council.

Wall, Tom. 2020. "Cramped living conditions may be accelerating UK spread of coronavirus." *The Observer*, 12 April 2020. Available at: https://www.theguardian.com/world/2020/apr/12/virus-hitting-hardest-modern-equivalent-victorian-slums

12

Chapter Twelve

Fire, acoustics, and services

Fire, acoustics, and services

Guy Marriage

DOI: 10.4324/9781003123873-12

Introduction

Two crucial but often overlooked aspects of apartment buildings are acoustics and fire compartmentation. Unlike an office building where openness and flexibility are key, apartment inter-tenancy walls do not move: neither for sound nor fire. Residents in a residential building will be far more aware of sounds and far less tolerant of them, especially while they are trying to sleep. Silence (or at least a significant reduction in noise) is crucial to an apartment dweller, but sleeping residents are more vulnerable to fires in buildings, as the brain does not respond to certain senses while asleep. Threading piped and electrical services through those inter-tenancy floors and walls is an unglamorous job but it is a high priority for the quality and safety of the completed building. Constructing an acoustically solid, reliable fire-resisting construction is a vital ability to develop.

Fire compartments

As every apartment will have its own legal certificate of title, then each apartment must therefore be a legally self-contained fire compartment. No fire should break out of this apartment into another, nor should a fire in any neighbouring units be able to break in. Floor construction and wall construction are crucial and junctions between them are just as important. While solving acoustic issues is important in terms of meeting the building code, stopping fire is an issue of human safety and so fire and structure take precedence over all else (Figure 12.1).

Floors – fire issues

A fire compartment must have a fire-proof floor below and above. Usually this is achieved by a concrete floor around 100mm thick, which will achieve a minimum of one hour fire-rated resistance (FRR 60/60/60) or even more. Any penetrations through this floor need to be carefully detailed and fire-sealed to avoid any issues.

CLT can also act as floor panels in a timber building. Mass timber functions well as a material in a fire, as the natural charring of the timber resists the timber burning right through and will stop the fire after the outer approx. 20mm is consumed. However, CLT often requires an extra layer of concrete topping, which improves fire-resistance and helps deaden sound transmission. CLT may also require layers of plasterboard underneath the floor to protect from the hottest part of any fire (KLH, n.d.).

Floors – acoustic issues

The floors constitute the primary means of resistance to acoustic issues between levels. Floors are highly susceptible to direct contact acoustic transmission. People talking or even shouting in the apartment above

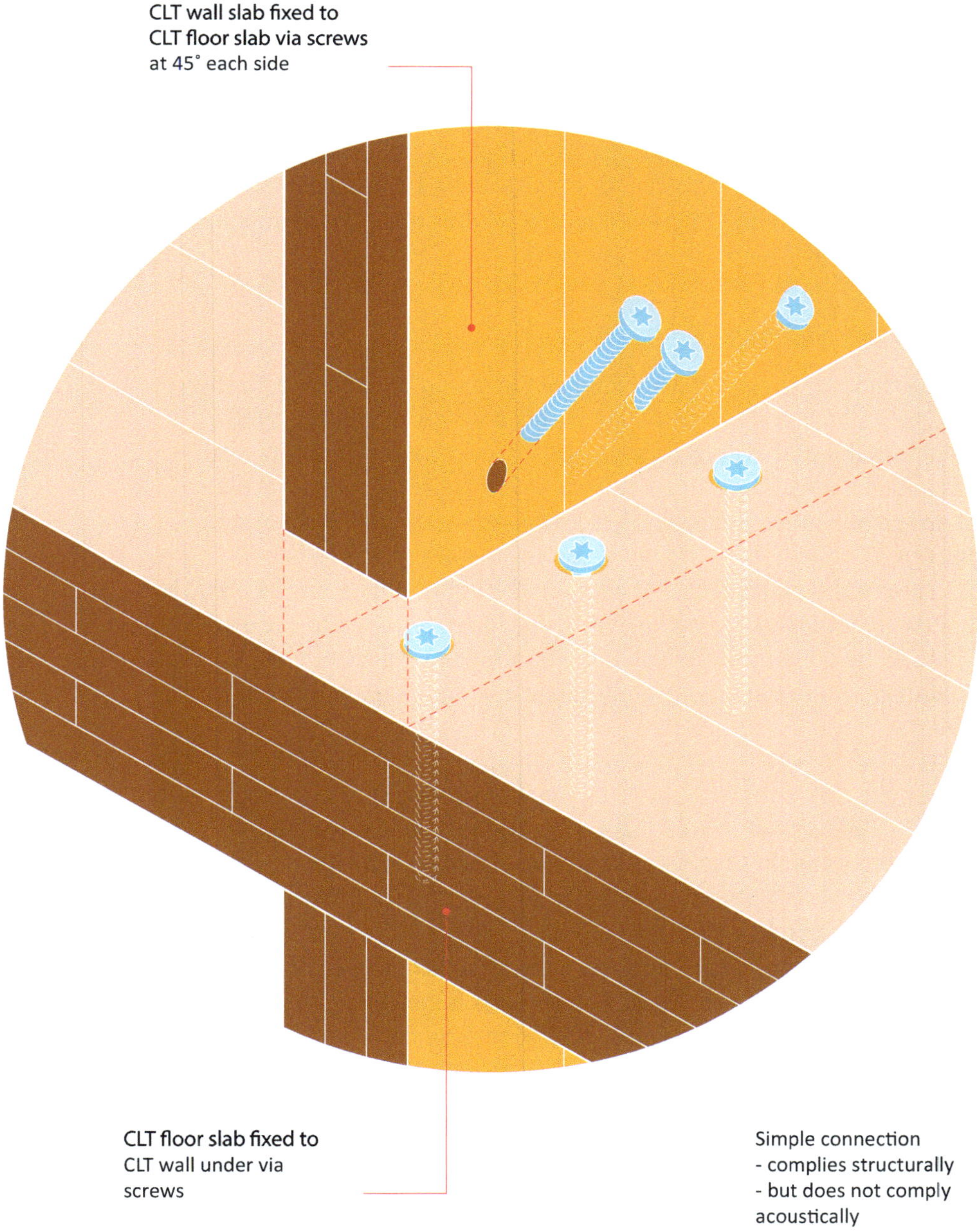

Figure 12.1 Screw-fixed internal CLT wall/floor junction.
Source: Hayes/Marriage.

should be incapable of being heard in the apartment below, but people simply walking around in hard-soled shoes can often easily be heard: the contact of each footfall transfers straight through to the floor below. Soft-soled shoes and slippers will transmit less noise than hard-soled footwear like boots and especially stiletto heels. It is up to the designer of the building to try and stop the acoustic transmission and this is best achieved at source. Acoustic control through floors depends on the construction materials chosen and their correct assembly.

Floors – noise reduction

A concrete floor on its own will not stop all the sound transmitting through the slab, so the floor will need treatment above as well as below. With a concrete floor the best way to deaden any direct impact sound is to specify an acoustic baffle. This can be as simple as a good thick layer of wool carpet on top of a thick rubber underlay. This will really muffle the noise sources but, of course, this only works where the floor surface is carpeted. In areas like bathrooms and kitchens there will not be any carpet, so the floor system (and ceiling below) needs to be designed to cope with this. This can be done by breaking up the floor/ceiling construction into layers, with vibration-absorbing separations and muffling acoustic baffles in between.

When the floor slab is mass timber CLT, there is another factor to think about. The lighter weight, lower mass, and high tensile composition of the CLT panel means that the mass timber floor can be strongly resonant, i.e. can transmit sound. This means that a building with CLT flooring can transmit footfall noise downwards quite well – it can be noisy and must be treated. For this reason, multi-storey apartment buildings require a layer of concrete topping screed over the CLT, around 40–50mm thick, to provide a deadening mass, lower the resonance, and reduce the sound transmission. It is convenient that this topping layer need not be on the critical path and so can be added after the CLT is installed, in order to not impede the building process. Similarly, layers of plasterboard can be added later to the underside of the CLT, not affecting the critical path of the building programme (Figure 12.2).

Walls – fire issues

Walls may be part of the load-bearing structural system or they may be non-structural infill. They need to be structurally integral from each floor to the underside of the floor above. It is crucial to understand the primary purpose of each wall, in order to detail the construction system correctly. Between the floors above and below, fire walls are needed, along every legal boundary. These inter-tenancy boundary walls (in the USA, called demising walls) need to be solid, safe walls in case of fires, usually

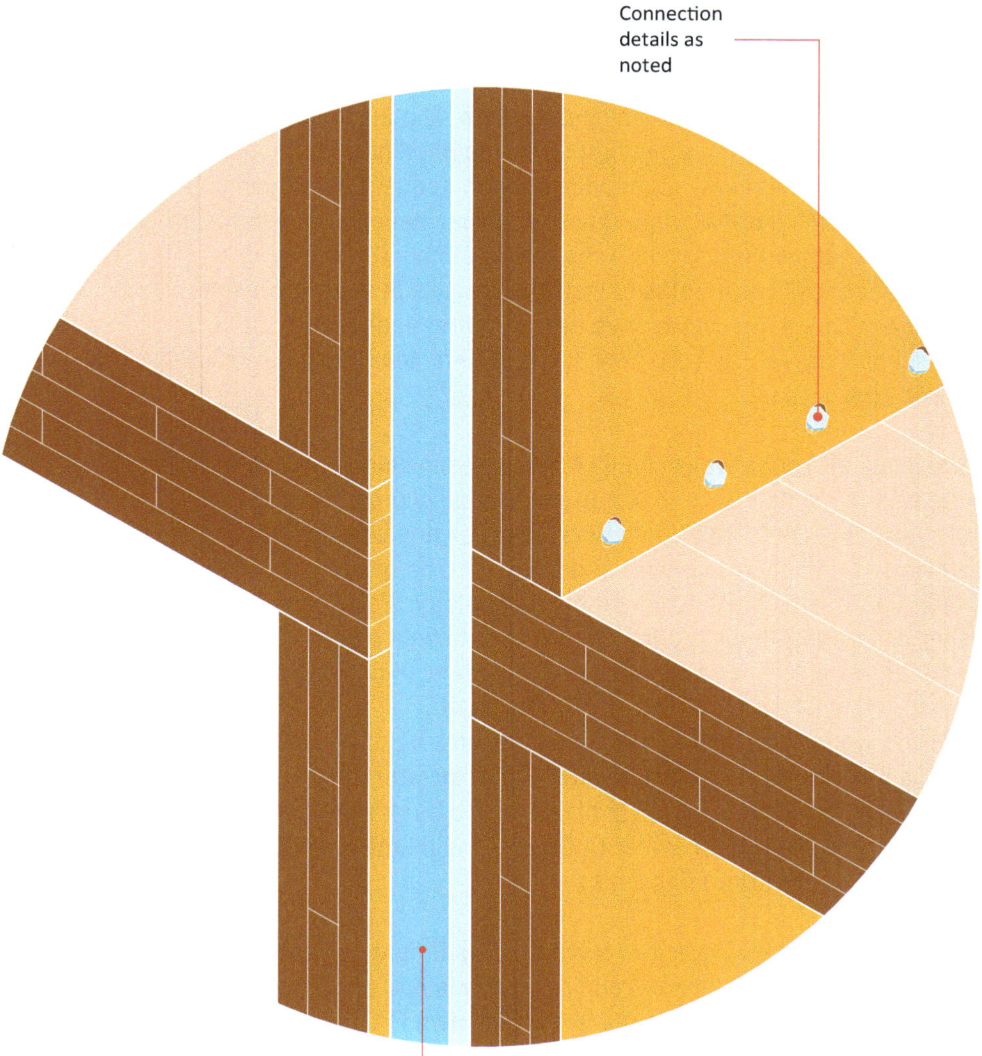

Figure 12.2 Acoustic separation CLT wall/floor junction.
Source: Hayes/Marriage.

achieving a minimum of one hour fire-rated resistance (FRR 60/60/60). This rating can be easily achieved with some materials such as brick, block, or concrete – a precast or in-situ concrete wall 100mm (4″) thick can withstand the heat from a fire for an hour or more. The thickness of the concrete/brick/block wall is dependent on the height it has to span, but a simple thin concrete wall will not provide enough acoustic reduction on its own. While (in the past) a single skin of wall may have been acceptable (especially when the boundary wall was a solid 215mm (8″) brick), it is now more acoustically preferable to install twin walls at every inter-tenancy boundary. Due to the destructive nature of fire, each wall must be fully self-supporting, i.e. from the floor to the underside of the floor above (Alread and Leslie, 2007).

Concrete block and brick walls should not be used in seismic-prone areas, however, as they could be prone to out-of-plane displacement in the case of a seismic event. They could wobble, crack and even collapse in a severe event. Both concrete blocks and bricks can be reinforced to withstand lateral forces, but this is old-school technology. Slow to construct and using wet trades, these technologies do not have a place in a modern apartment building. We need speedy, reliable means of construction.

CLT can form walls in many different forms: an inter-tenancy wall, a structural wall, an externally-insulated façade wall or an internal wall. The CLT will typically require double layers of plasterboard each side to form the fire wall – or may also require wider CLT to cope with any potential charring in the case of fire (Stora Enso, n.d).

Other materials such as light timber framing (LTF) or light gauge steel framing (LSF) can be used but will require double layers of gypsum plasterboard (drywall) to each side to reach the minimum FRR 60/60/60 level required. Gypsum plaster does not burn, so a thin layer can have considerable effectiveness as a fire-retardant. Sheets are laid in a staggered pattern so that the joints between sheets do not align. The name 'drywall' describes it well: it is a wall which can be installed dry and must be kept dry as the gypsum plaster swells and degrades in contact with water. For internal walls within the apartment such as bedroom walls, this construction system may be perfectly adequate with a single layer of plasterboard each side. Your apartment building is therefore likely to have a number of different wall construction systems. Understanding the construction junctions between these wall systems is crucial to the detailing (Figure 12.3).

Walls – acoustic issues

The acoustics of walls is also crucial. A solid mass wall will resist airborne sound better than a hollow wall, although both wall systems are susceptible to direct impact noise transfer. Sounds in the acoustic vocal

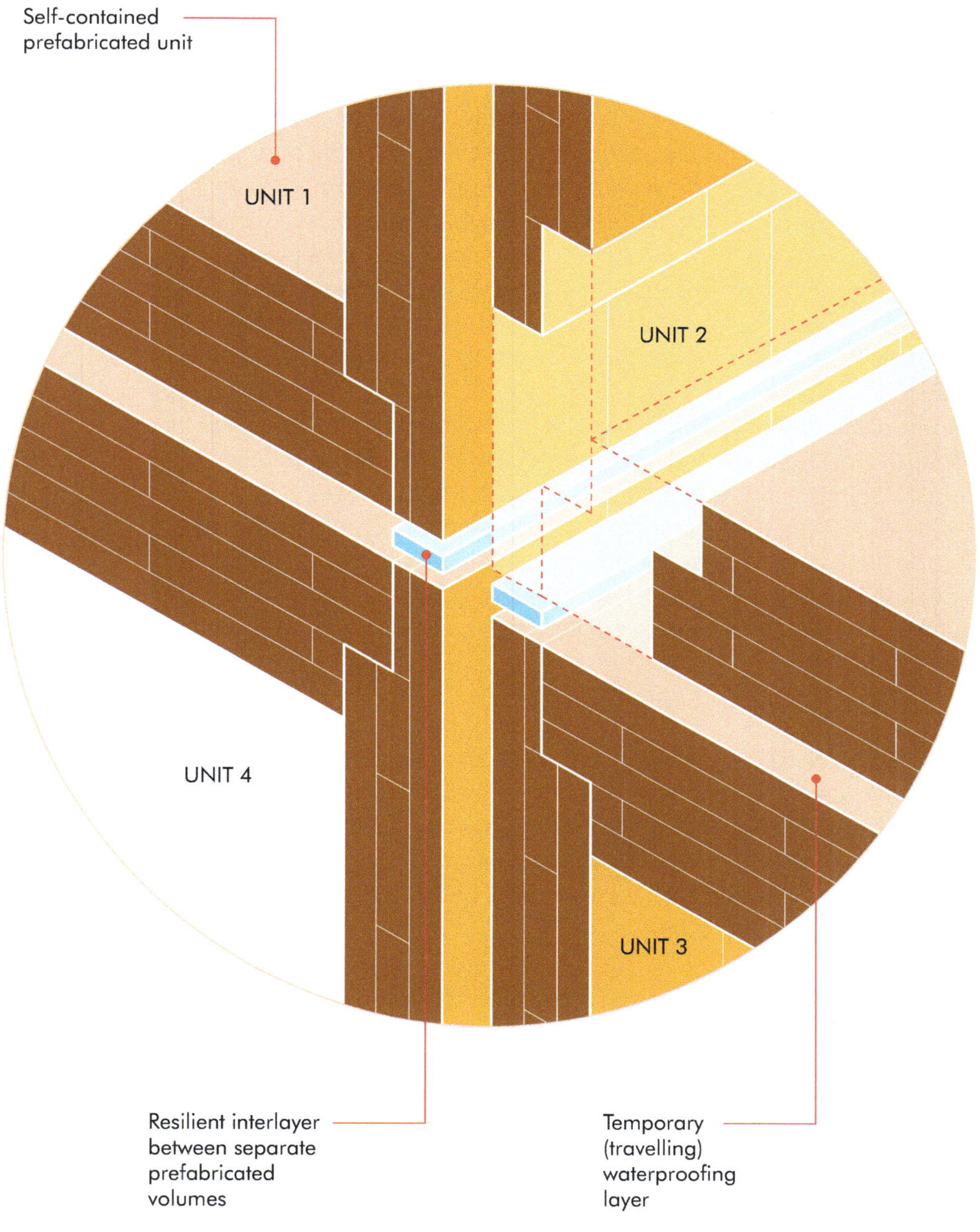

Figure 12.3 Multiple module CLT wall/floor junction.
Source: Hayes/Marriage.

range (i.e. someone talking, yelling, or playing music) can also be heard through a wall, usually slightly muffled. Direct impact noise can be as simple as someone tapping on the other side of the wall and the noise transferring straight through. Noises can transfer into a wall from a connecting floor – footsteps next to a boundary (inter-tenancy) wall can be highly audible, especially if the floor is not carpeted. Running water and flushing toilet sounds can also readily telegraph through the structure and make themselves heard next door or on the next floor down. Events as simple as the click of a light switch can sound irritatingly loud to someone on the other side of the wall, especially in the middle of the night. In Germany, generations of apartment dwelling has made residents sensitive to sound and excess transmission of noise is actually illegal outside the specified hours from 10 pm to 6 am. Room volume regulations mean that no noises can escape from the apartment (Bundesministerium der Justiz und für Verbraucherschutz, n.d.). Transgressions may be punished with a fine of up to 5,000 euros, and even personal hygiene actions such as showers are asked to be kept short; woe betide you if you want to practice your trumpet for more than two hours a day (and even less on Sundays). All of these issues need addressing in order to construct an audibly comfortable wall.

Walls – multi-layer construction

A single-skinned wall (LTF or LSF) is simply not acceptable in any form as an inter-tenancy wall. For acoustic reasons, inter-tenancy walls are often composed of several different layers, with vibration-free separation in between. This typically means that a twin-wall system is required. While this may mean that the material cost of the wall construction has doubled, it is necessary for people's sanity.

Two main designs are commonly used, usually depending on the location of the insulation layer. In the first scenario, two layers of linings are installed on the external side of each wall and supported by steel or timber studs, with insulation being installed in between the two ranges of studs. The insulation stops at the column on which battens are fixed to, supporting the linings and ensuring visual continuity of the walls (Figure 12.4).

In the second scenario, the layout changes so that two layers of insulation are installed in between the studs, and a gap is created at the centre of the twin walls, with sheathing installed over the two ranges of studs. Depending on the dimensions of the column, the layers of insulation continue from either side of the structure or stop when reaching it. In this case, simple battens against the column and supporting the linings can be used.

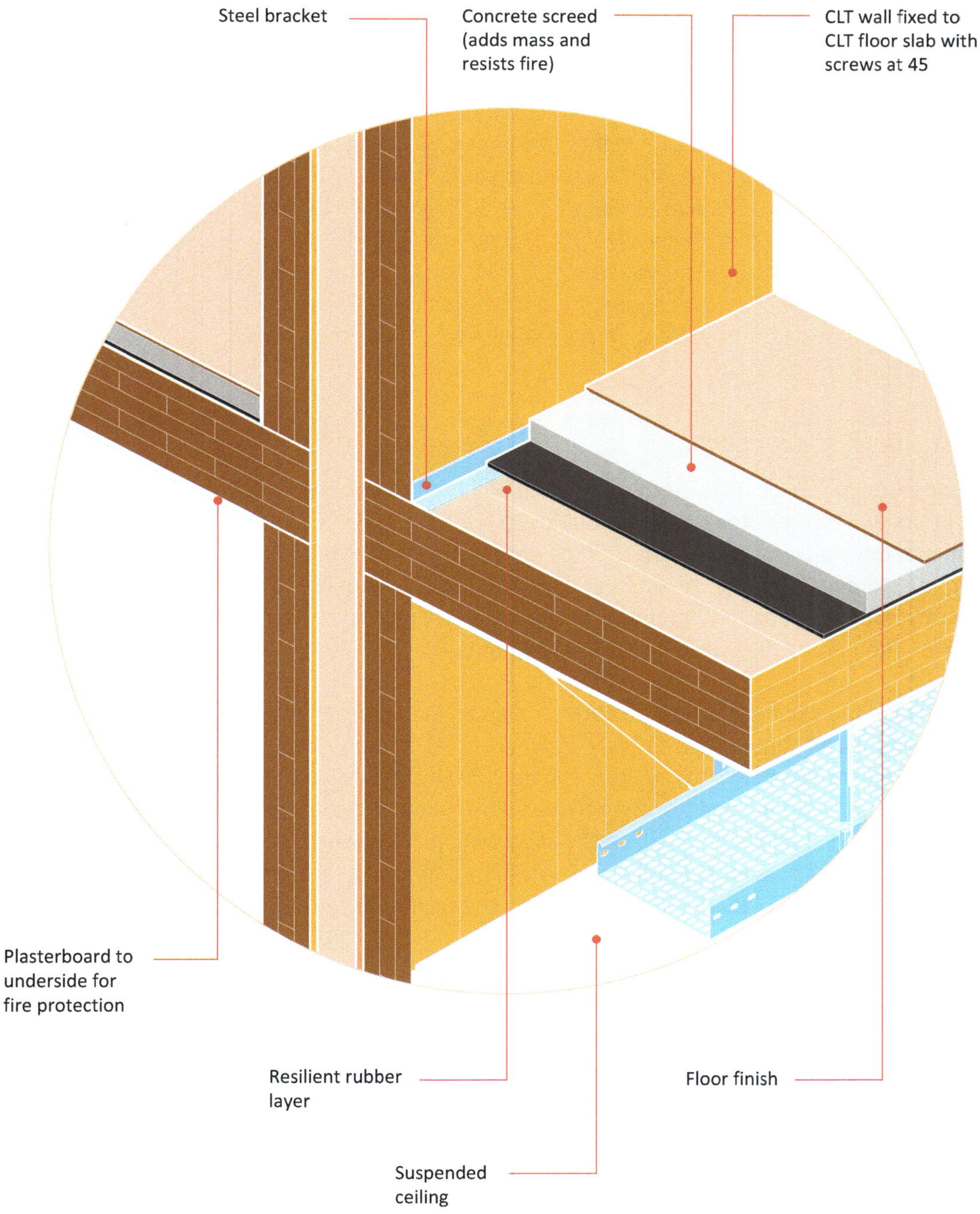

Figure 12.4 Full build-up of CLT wall/floor junction.
Source: Hayes/Marriage.

Inter-tenancy walls can also be totally independent from the building's primary structure and therefore not aligned along the structural grid. As non-bearing elements, their composition and layout remain very similar to the two previous scenarios presented but with no columns. The use of twin walls ensures a structural independency between the two walls, therefore avoiding any direct sound transmission. To have any chance of reaching the desired acoustic level, there should be no connections between the walls. To further improve this feature, steel resilient channels can be added. These elements are fixed horizontally to the studs and support the linings while isolating them from the framework. The channel attenuates vibration transfer through the walls as it reduces the critical noise frequency (Table 12.1).

Walls – head and foot details

Construction of a successful acoustic barrier will require attention not just to the wall system itself, but particularly to the details at the top and bottom of the wall. The entire ability of the wall can succeed or fail on the quality of the acoustic/fire detail at the base and the head of the wall. Ensure you understand the functional priority of every junction you detail – stopping fire spread must always take precedence.

These areas are small but crucial to detail correctly. At the foot of any inter-tenancy wall, there needs to be an acoustic break to stop vibrations in the wall from transmitting into the floor. The acoustic material also needs to be fire resistant: you will need a strip of fire-rated sealant at the underside of the wall, or at the base of the plasterboard/drywall. This will help break up any audio transmission and will also retain fire compartment integrity. *Robust Details* has an excellent set of details you should refer to (2020).

The head detail is similar but has one crucial difference: the designer needs to allow for sag. Any floor slab, if the span is long enough, can sag very slightly over time. A concrete slab floor can, and will, sag in the centre of the span, depending on its own self-weight (dead load) and even a CLT slab can do the same. Your engineer will advise on the thickness of the slab and the span between the columns or supporting walls, but regardless, the head detail needs to be able to absorb any flex in the floor slab above without creating cracking in the wallboard below. In seismic locations this detail becomes even more important, as the detail will need to flex as the floor slab vibrates up and down and the building rocks from side to side during an earthquake. It is difficult to get a detail that can both stay acoustically and fire-resistant intact, while also allowing movement, so specify a system with a proven track record. A seismic head joint can be composed by allowing the wall to slide inside a tight-fitting sleeve over the top, while acoustic/fire sealants at these points are crucial (Figure 12.5).

Table 12.1 Floors and walls – fire, acoustics and construction issues

	Floors	Walls
Fire-rated construction	• 100mm (4″) of concrete slab will achieve minimum FRR 60/60/60. • 40mm (1.5″) of concrete topping will provide FRR 60/60/60 to a CLT floor, in association with: • Two layers of plasterboard to the u/s of exposed CLT slabs to stop the spread of flame. • Fire-protect all exposed steel beams, columns, and connection nodes supporting any floor.	• 100mm (4″) of concrete wall will provide minimum FRR 60/60/60. • Fire-rate external faces of inter-tenancy walls with double layers of gypsum plasterboard each side, stagger the joints. • Fire-stop all gaps between fire compartments. • CLT walls will require cladding with gypsum plasterboard to achieve fire rating; also helps with acoustics.
	Floors	**Walls**
Air-borne sound	• Install lowered ceiling with acoustic tiles/plasterboard, with acoustic batts insulation on top. • Run water/sewerage services in a separate riser. • Do not run plumbing and drainage from one apartment through the ceiling space of another apartment. • Run services horizontally only through raised floors, or in common space of the ceiling below, i.e. over corridors.	• Mount plasterboard on resilient channels, i.e. vibration-free substrate. • Design in void space between twin walls. • Stagger joint spacings on either side of the wall. • Infill framing with acoustic batts to help absorb any sounds, or thread a continuous layer of acoustic batts between walls.
	Floors	**Walls**
Direct-impact sound	• Avoid bare hard surfaces such as concrete. • Specify carpet on underlay. • Install resilient layers between adjoining hard surfaces. • Use a hard rubber mount between adjoining CLT slabs.	• Use multi-layer construction: install twin walls for inter-tenancy, mounted independently so that they do not touch. • Do not install any light-switches or power-points in inter-tenancy walls. • Heavy batts insulation needed in between walls. • Acoustic and fire-rated sealant to be applied at both the top and bottom of inter-tenancy walls.

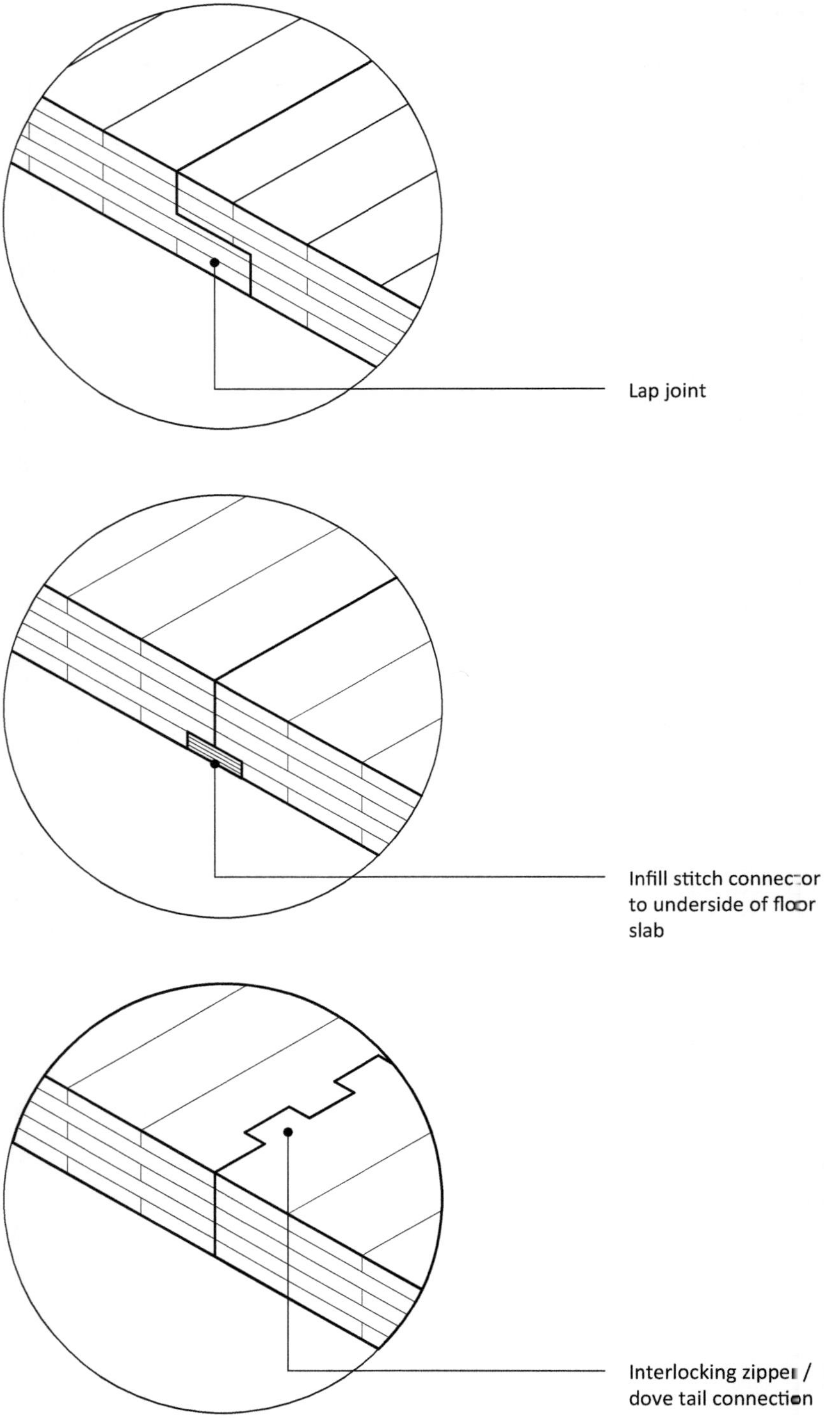

Figure 12.5 CLT floor panel lapping joints.
Source: Hayes/Marriage.

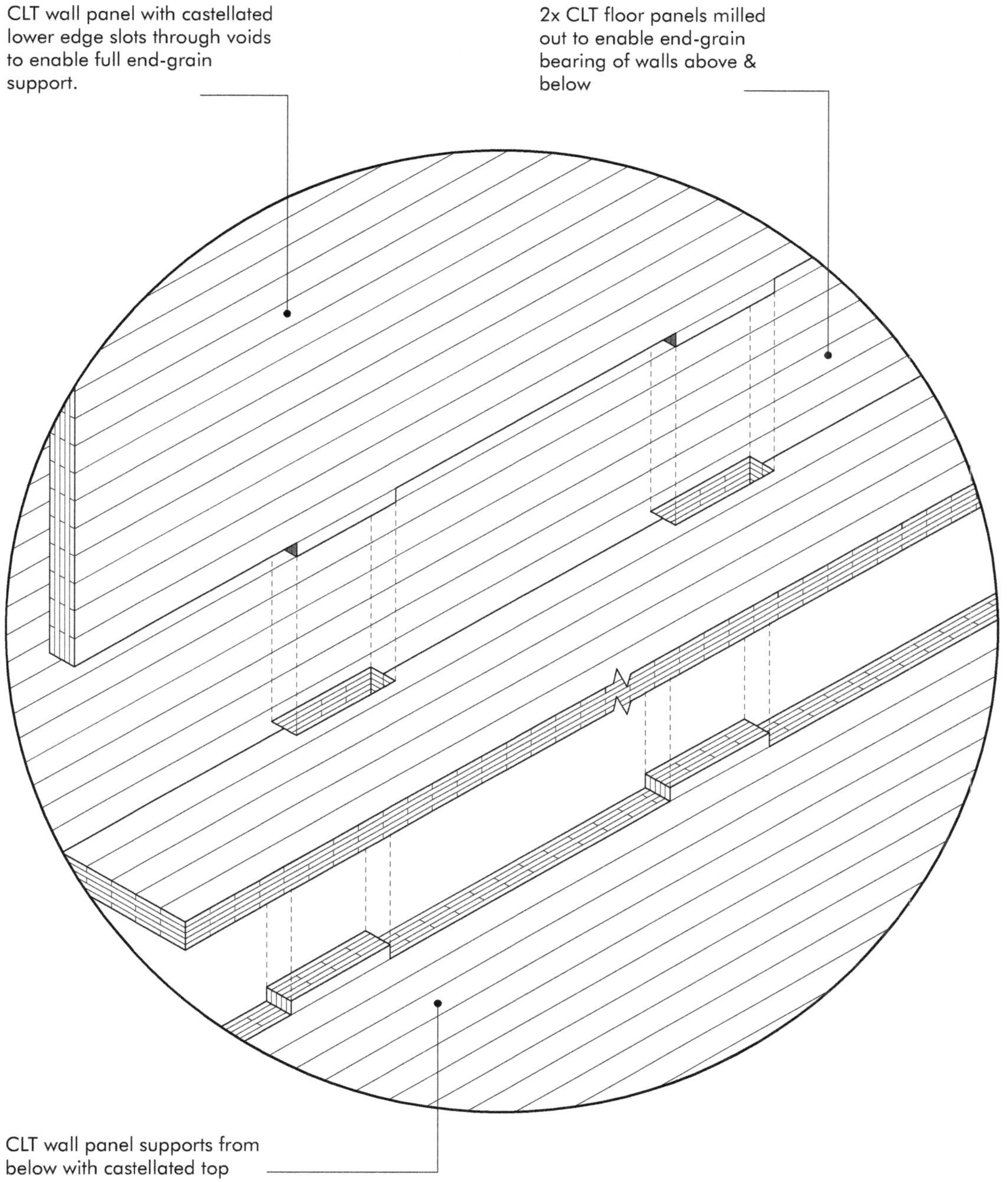

Figure 12.6 CLT wall to floor castellated slot connection.
Source: Hayes/Marriage.

Ceilings

The construction of the ceiling is an integral part of the acoustic solution – and sometimes even part of the fire-resistance solution as well. It is important to note the order of construction – fire-resistant walls must go all the way up to the underside of the fire compartment floor slab above. If a concrete slab is used, no further action is needed: the wall can be built right up to the underside of the slab above. However, if the floor slab is built from CLT, and requires a layer or two of plasterboard, this must be applied to the underside of the CLT before any internal fitout wall construction commences.

Hang the ceiling from the floor above via an isolated structure, i.e. hanging by wires with resilient channels to reduce acoustic transmission. The ceilings will be composed of one or two layers of plasterboard, and will usually be backed up with a thick layer of acoustic batts insulation. It is vital not to have any penetrations through the ceiling that might subtract from the efficiency of the ceiling construction. This means that lights should not be recessed into the ceiling, but should be surface mounted, or specifically designed to cope with the construction. The routing of services through and above the ceiling must also be taken carefully into account (Figure 12.6).

Services

Plumbing and drainage routes need to be thought out early on, as does the means of electrical power distribution. As with commercial office buildings, all services should be accessible from within each floor. Do NOT plan to run drainage from one apartment through the ceiling space of the apartment below, although you can run it through the ceiling of the common areas on the floor below. Keep all services under a false floor or in a self-contained suspended ceiling. Nor should you plan to run electrical wiring or plumbing within the inter-tenancy wall: run services in non-structural walls leading to a common vertical services riser. Install a floor drain to all bathrooms/laundries in case of flooding. Carefully plan out services routes early on in the design, so that you can avoid any expensive, awkward, or unfeasible crossovers of different services in the ceiling. Don't leave the services reticulation till last – start early!

Services riser

Position services risers early on where they can be easily accessed from the common area (i.e. corridors or core) and ensure that they just go straight up and straight down the building. This will make the building contractor's job much easier and will greatly aid in any future maintenance: it also makes less noise. A simple solution is to mirror adjacent apartments so that one vertical services riser can be shared by both apartments. If the services riser can be situated in the circulation/structural core, this will also assist with the flexibility of the apartment layout, but be aware that this can increase the length of the services that run to the core. For this reason, you may often have a slightly lower ceiling level in the core or corridors than is achieved in the apartment itself.

Services within apartments

In Europe, it has been common practice to physically chase a groove into a solid brick inter-tenancy wall for inserting a wiring conduit and then plastering back over it, but this is outdated thinking as it is difficult to upgrade later. Avoid any wiring within an inter-tenancy wall at all: this avoids the issue of noise transference through the light switch or plug socket, as well as avoiding issues with fire walls. Plan, therefore, for wiring to be installed on internal fitout walls only and note that in a multi-storey building the wiring is most likely to descend down the wall from the suspended ceiling above or from below if a raised floor is installed (Figure 12.7).

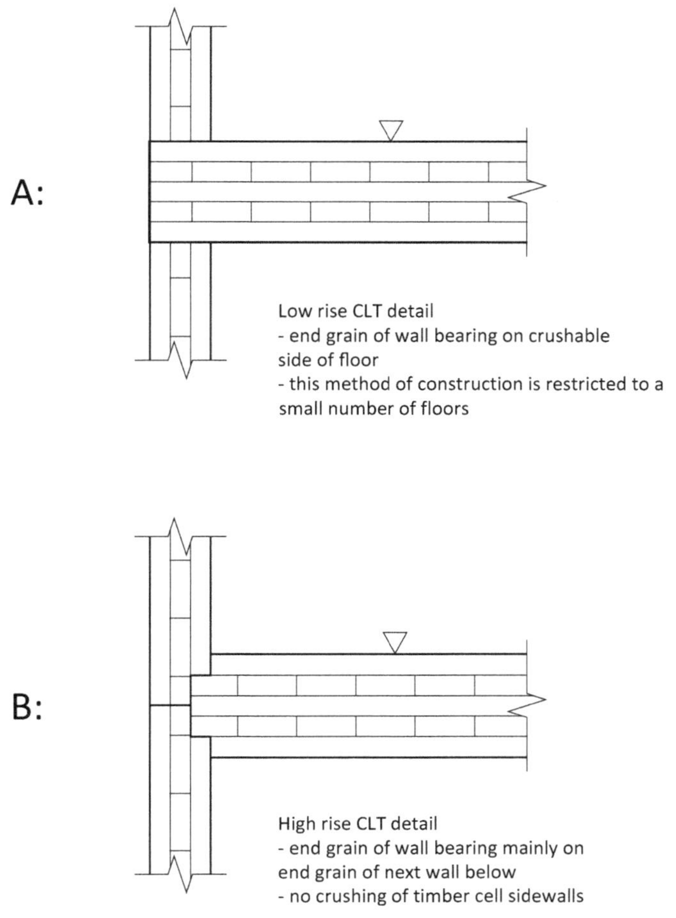

Figure 12.7 Low-rise/high-rise CLT details.
Source: Hayes/Marriage.

Sprinklers within apartment buildings

Sprinklers save lives. All multi-storey apartment buildings should have a sprinkler system built into the base build. The added cost at the start for the developer is soon paid back to the resident by means of the added value, peace of mind, and hopefully lower insurance premiums. Sprinkler layouts must conform to a code and so are installed only by registered installers. So that the results are not all over the place, the architect should be involved in the setting out of sprinkler positions as well as all the other services: ventilation grilles, fire alarm sounders, and the position of lighting. Trying to align these services is simple on paper but more difficult in reality – sprinkler positions and the sprinkler supply pipes rule supreme. Lighting, ducting, and other services must be routed to meet the fixed position of the sprinkler head.

It is worth mentioning this once more: the lack of sprinklers in the Grenfell Tower building fire in London caused 72 deaths and untold heartache as well as millions of pounds of damage. A far simpler practice would have been to ensure that all apartment tower blocks, new and old, are fitted with sprinklers for the safety of all.

Works cited

Alread, Jason and Leslie, Thomas. 2007. Design-Tech: *Building Science for Architects. Oxford,* MA: Elsevier.

Bundesministerium der Justiz und für Verbraucherschutz. n.d. *Gesetz über Ordnungswidrigkeiten (OWiG) 117Unzulässiger Lärm.* Germany: Bundesamt für Justiz. Available at: https://www.gesetze-im-internet.de/owig_1968/__117.html

KLH. n.d. *Panel Types, Surfaces &Technical Characteristics.* Available at: https://www.klh.at/en/panel-types-surfaces-technical-details/#technical-details

Robust Details. 2020. *Part E Robust Details October 2020 Revision.* Available at: https://www.robustdetails.com

Stora Enso Wood Products. n.d. *Stora Enso CLT Technical Brochure.* Available at: https://www.storaenso.com/en/products/wood-products/massive-wood-construction/clt

13

Chapter Thirteen

Façades and cladding

Façades and cladding

Guy Marriage and Nabil Jose Allaf

Façades and cladding

DOI: 10.4324/9781003123873-13

Introduction

Your high-rise apartment building will have an external wall system composed of a mixture of glazing and solid walls. It may be flat or deeply modelled in three dimensions and it may have extensive balconies to all façades, or none at all. The external façade is an exercise in proportions, aesthetics, and function as well as overall appearance. Some building façades are integrated with the structure, especially in the case of masonry or in-situ concrete fire walls. Too often the building will be judged more on what it looks like, than what happens inside. The façade mitigates the external environment and is therefore doubly important, being valued from both inside and from outside. Some would argue that it is the most important part of the entire building. What it looks like is merely one part of the equation.

Environmental considerations

Firstly, the façade must be able to breathe. Human beings take in food, oxygen, and water and emit water vapour and carbon dioxide via breathing. A building is in many ways similar, with a skin that needs to breathe. We simply can't have a fully sealed façade like a plastic bubble as the humidity level would build up inside to unhealthy levels. We need a steady supply of fresh air containing oxygen and the building façade must be designed to allow the water vapour to exit out, even in cold climates where the windows may be closed. Insulate the building well to keep the warmth in during winter and keep excess heat out in summer. Install vapour barriers to stop any water vapour condensing into water droplets within the wall build-up. The wall construction system will differ depending on the predominant climate where your building is (Donn and Thomas, 2001). The construction answer is not simple or monolithic, as any external wall system will be composed of a number of layers including structure, weatherproofing, insulation, ventilated cavity, internal lining and, of course, external cladding (Ramsdonk, 2020).

Cladding material choices

The choice of materials available for your cladding is wide; however, you cannot use the CLT or LVL panels as the final external cladding. CLT works well as a structural solution to the external wall but the glues and materials are simply not suitably waterproof for longevity. CLT needs further layers: a vapour barrier, considerable insulation over the external face (much better than on the internal face) and a reliable weatherproof yet breathable external cladding system that encloses the entire external façade. This requires careful thought and planning and is very reliant on the climate of the site. Sutherland looks at the wide variety of cladding materials in *Tall* (Marriage, 2019), along with all their positive and negative attributes.

Non-flammable cladding

Firstly, make absolutely sure that the external cladding is constructed from an inert material and cannot create any fire problem – certainly not any form of plastic! A solid form of metal panel is often the most appropriates – zinc-coated profiled steel is certainly a valid option, as are terracotta tiles, stainless steel, enamelled steel sheet, brick facings, solid aluminium siding, concrete, brick, and glass. Do not think of materials in isolation, but design and specify a fully-coordinated cladding system. The following section looks in more detail at some of the materials commonly used.

Brick

A traditional brick masonry cladding is built in-situ, brick by brick, with mortar fixing each brick in place. Construction is therefore solid, weatherproof, reliably long-lasting, but very slow to construct, and often the base of a traditional brick masonry wall is many layers thick. Bricks are created from kiln-dried clay, so have a natural variation in colour and texture that can add a lot of character to the building, plus the thermal mass of brick masonry can smooth out heat gains and losses within the building. Twin skins of brick cladding have been common in many places in the past, featuring an insulated void behind the bricks with restraining clips integrated into the structure of the main façade construction. However, brick cladding installed in traditional mortared fashion is completely unsuitable for use in seismically-active countries like New Zealand, Japan, Chile, or regions like California. Unreinforced brick masonry buildings are not your friend in an earthquake.

For multi-storey high-rise buildings, bricks can be more reliably installed on a floor-by-floor basis as part of a prefabricated system. Concrete precast panels may be finished with brick facing 'slips' so that the building appears more brick than concrete. The bricks are laid face down in moulds and the concrete is poured in to bind them from behind, or the bricks may be fixed with epoxy glue to the concrete panel at a later state. The finished building has the appearance of a brick façade but has the stability of a precast panel system. The precast panels will then need insulating internally to achieve a suitable thermal rating, usually necessitating a secondary wall internally, or at least strap-and-lined insulation on the inside face of the walls.

Terracotta

Terracotta means 'baked earth' and is a more graceful form of fired earthen tile. Colour variations of the source clay are reduced but not eliminated. The key advantage of modern terracotta is that it is a complete cladding system. Terracotta panels are extruded from a sloppy clay mix into elaborate double-skinned shapes which encompass weight-saving voids and integral grooves for hanging from clips fixed to the building. Like bricks, the porous surface means that the panels will absorb some water and slowly dry out, but both bricks and terracotta are also available with a surface glazing which sheds most of the water. Unlike bricks, there is no mortar fixing the units together: modern terracotta systems are open joint, free-draining, and the individual panels are simple to replace. This makes them ideal for seismic areas, as the building moves and flexes during an earthquake, the open joints can easily cope with those movements. As with any open-joint system, the actual weatherproofing layer lies behind the decorative cover.

For an even more colourful version of this form of cladding, fully glazed ceramic tiles can be used, giving the building a colour palette previously not available. A bold example of this is the St Giles Circus development (architect: Renzo Piano, 2010). Here, swathes of colour in orange, lemon and lime brighten and startle London's more traditionally subdued colour range. Colours aside, the ceramic panels themselves are a tour-de-force of prefabrication, carefully clipped together to form a richly detailed and unified façade system.

Concrete

Bare concrete as an external façade material has the rugged reputation of never needing further maintenance – but this is largely untrue. Water will stain concrete, and water will cause damage over time. As concrete is dense, but porous, it can absorb water. Generally, it will dry out again satisfactorily, but chloride salts within the concrete can migrate through the mass with continued wet/dry cycles. Reinforced concrete (RC) contains steel and this steel will rust over time as water penetrates inwards. As the rusting steel expands, the concrete cracks and spalls, seen to horrible effect in Champlain Towers South condominiums, Surfside, Miami (2020).

If you really want the 'Béton brut' bare concrete look, be prepared for endless arguments about brutalism. Examples can be used both as reasons for and against – look no further than the Barbican Estate in London (architect: Chamberlin et al., 1965–1976), which satisfies both sides. Some love it, some hate it. Typically, architects tend to be the ones who love it, and sadly, the general public tend to be in the hating majority. In reality, the massive thermal mass, excellent detailing and good maintenance programme mean that the Barbican is a very comfortable building to live in. The thick concrete walls slow down the loss of heat in winter and provide a similar time delay against hot summer days.

Concrete precast panels will tend to be insulated on the inside for reasons of durability, but as you will know, the best place for insulation is always on the <u>outside</u> of the building structure and this goes for concrete as much as anything else (Donn and Thomas, 2001).

Metal cladding

Steelwork to the exterior of a steel-framed building may often be finished by a cladding that is metallic in nature, so that it retains the look of a metal building. Because steel potentially has issues such as rusting, thermal bridging (heat transmission), thermal expansion and a loss of strength under fire conditions, it nearly always has an external separate cladding used, such as a clip-on extrusion that covers the rustable, fire-protected steelwork below. This is likely to be another material such as stainless steel, zinc-coated steel, aluminium, or aluminium composite panels. Metal cladding has great credentials – it's very strong even when used as a thin surface and it can be a relatively lightweight cladding. The jointing system is key; as metals expand and contract significantly under heat loads, the method of fixings needs to be able to cope with movement. On high-rise buildings, these fixings are usually stainless-steel clip systems.

Glass façade cladding and privacy

In a similar manner to commercial office buildings, apartment buildings can also be clad in a fully glazed curtain wall. As a material, glass is great for creating a smooth, clean, waterproof façade system, but it does raise an important additional issue: privacy. Intimacy is rarely a key criterion for office and commercial buildings, yet is highly relevant when it comes to the design of the façade of tall residential buildings. While living in a glass-clad building would give the users the luxury of unique outside views, controlling the visual and acoustic surrounding environment is compulsory in order to maintain a certain level of privacy. This control directly influences the degree of interaction with the outside and therefore the design of the façade. Careful consideration for the design of the glazed façade is necessary to ensure there are proper sound and thermal barriers, as well as appropriate visual barriers in places.

This can be put in contrast with office and commercial buildings which commonly display fully glazed façades, directly originating from the 1930s International Style yet still carrying the notion of modernity and excellence. In these buildings, the relationship with the outside is sought. The glazed façade provides full visual freedom while establishing a definitive physical boundary. The building occupants can see the people on the street and the latter can, to a certain extent, look inside the building.

Two main types of curtain wall cladding systems can be used by the architect.

Firstly the ***stick*** system refers to a self-supporting aluminium frame made of a combination of vertical mullions and horizontal transoms supporting the windows. In combination with the glazed areas, opaque panels can be added, commonly in front of the floor slab at each level. These spandrel panels can vary in depth, thus allowing the architect to adjust the outside views and amount of natural light to their desire. This system is particularly appreciated for its long continuous horizontal glass bands at each floor. Often, in a fully glazed façade, the spandrel may be back-painted glass or finished with a ceramic frit coating to disguise the slab, with insulation and mechanical fixings behind.

The alternative to cladding system is the ***unitised curtain wall***. This system uses prefabricated wall panels built up off-site, which ensures optimum manufacturing quality control and speed of installation once delivered. They are commonly one storey high and one or more modules wide, depending on the physical limitation of the transportation shipping them to site. The unitised panels are often hybrids meaning that they are made of several layers of different materials which form a fully complete envelope. The combination of the envelope can vary depending on the architect's requirements but can include components such as internal linings, insulation, concrete precast panel, aluminium frames, and glass. These types of wall panels are appreciated for the variety of pattern they can generate depending on their geometry, volume, and size.

For residential buildings, three main approaches with different degrees of privacy can be considered. The first one, and most visually exposed, is identical to office and commercial buildings using a similar fully glazed design for the façades. It presupposes a fully sealed façade however, which may not be so desirable by the inhabitants of the residences.

The second approach uses balconies as horizontal privacy screens. This approach allows maintaining fully glazed façades while reducing the cone of vision from outside in. These architectural features can be expressed as projecting or recessed balconies.

The final approach, providing the highest degree of privacy, corresponds to a reduction of glazed surface. The goal is to use the cladding of the façade as a vertical screen and determine the required location and amount of glazed surface.

External thermal insulation composite systems (ETICS)

A very different system is ETICS, where an external insulation product is directly covered by a plastered finish. Both New Zealand and Canada are countries that are weary of this approach, having both been subject to severe and expensive issues with leaky buildings (Barrett, 1998; Bonner et al., 2020; Dyer, 2019). Also known as Externally Insulated Façade System (EIFS), the systems can work well where detailed appropriately, particularly with regard to the substrate. A solid concrete substrate will work well for adhesive-fixing an external insulation layer to, whereas a more flexible timber-based substrate (plywood or OSB) has a poor chance of making a successful cladding system. Reputable companies will promote their products and systems, with excellent insulation layers up to 360mm thick, tying the external plaster cladding to the insulation layer with silicate plasters fixing glass fibre mesh in place. The chief problem for any such system is movement of the substrate and the resultant cracking of the outer surface. Even a small crack can allow considerable water ingress, and while this may be easily spotted and fixed low down on a façade, it is a nightmare to locate and repair high up on a façade. Recommended only for low level cladding.

Another option is to couple an insulated layer with an over-cladding of another finish, but as has been noted with Grenfell House, this too has its challenges. One study noted that while 100% of façades without cavities pass the associated fire test, only 55% with cavities pass (Bonner et al., 2020). Façades with fire retardant ACP or with HPL panels performed worst.

Cavity barriers

Ensure that non-flammable, permanently-fixed cavity barriers adequately block off any chance of smoke or flame from rising vertically or spreading horizontally. Grenfell's issue with cavity barriers was that they were not only badly placed, but they also did not survive: barriers need to be permanent stops to the spread of flame and must also act to discharge water wherever possible (Bonner et al., 2020). Seek specialist advice from façade engineers to check your cladding drawings.

Over-cladding existing in high-rise buildings

Many of the rapidly-built tower buildings constructed mid last century had brick and concrete external facades that suffered badly from a lack of maintenance. Recently, a concerted effort has been made to over-clad them to keep the residents inside warm and dry, with external insulation, new windows and new weatherproof cladding. Grenfell Tower was one such high-rise where the over-cladding went tragically wrong on 14 June 2017. While the original 1974 brick and concrete cladding on Grenfell was porous and created cold, damp apartments, an extensive refurbishment in 2016 sought to resolve this by over-cladding first with a thick layer of insulation and then a layer of shiny new cladding on the outside of a drained and ventilated cavity. Windows were replaced with new plastic windows in a new position, closer to the outside of the building. The official inquiry into the Grenfell Tower fire has revealed that this approach was fundamentally flawed (Grenfell Tower Inquiry, 2019).

It turns out that virtually all of these newly installed elements were badly designed and poorly installed: internal plastic (PVC) window jambs melted and let a small fire from a mal-functioning kitchen fridge pass into the external cavity, so that the polyethylene inner to the aluminium composite panels (ACP) melted, dripping what can only be described as liquid fire onto panels below. The ACP itself caught fire, with the thin aluminium skin burning up, fuelled by the supposedly inert material within. The polyisocyanurate (PIR) polymer insulation charred and burned, while the cavity barriers did not adequately contain the spread of flame and instead funnelled the fire both up the building and sideways, into other residents' apartments (Grenfell Tower Inquiry, 2019). Other apartment buildings have since had their (similar) cladding removed, which in turn has led to other problems (Taylor, 2020), and only some have been successfully reclad (Murray, 2020). Clearly, cladding systems will need to be far better specified, detailed and installed in the future. The Torre dei Moro fire in Milan (August 2021) caused also by flammable cladding shows that we have a long way to go.

Insulation

Apartment buildings have the distinct advantage over individual, detached residences, that they are in effect insulated by their neighbours. The amount of exposed external cladding is small: most apartments are shielded from the elements by apartments on either side, as well as by the apartments on the floors above and below. If your apartment is a box with only one side exposed to the atmosphere, which means, that five sides are shielded by neighbours and this provides effective insulation. It is this hive effect that means that well-designed apartments are usually, warm, comfortable, and cost-effective to run. My old apartment in London was above a video games arcade and so we survived seven winters with zero heating, except for the excess heat rising through the 300-year-old floorboards. Of course, the corollary was that it was unbearably hot in summer…

Broadly and very simply: ensure that all insulation to any high rise is completely non-flammable. An inert insulation material that can be used is mineral wool. Being composed of rock that has been stretched and spun into a matrix of tiny fibres, mineral wool panels cannot physically burn. Obviously, avoid any mineral fibres that contain asbestos.

As the use of conventional insulation materials installed on the façade are not compatible with a fully glazed design, architects aim for alternatives. This has led to the increasing popularity of products like acoustic laminated glass as well as heat and glare reflective glass. These types of glass are noticeable for the mirror-like effect they can generate on the façade: they are cooling on the inside, but can also be alienating for those on the outside.

Alternatively, the use of double skin façades is also a relevant solution in terms of acoustic and thermal barriers. These façades are composed of two glazed layers separated by a naturally ventilated cavity, but this technique is more common for commercial buildings than for residential. This can be explained by the cost of building up two layers of façade as well as the increased cost of cleaning and maintenance. Residential buildings are not as hot as offices due to the reduced number of people and not as much technology that generates heat, such as computers (Figure 13.1).

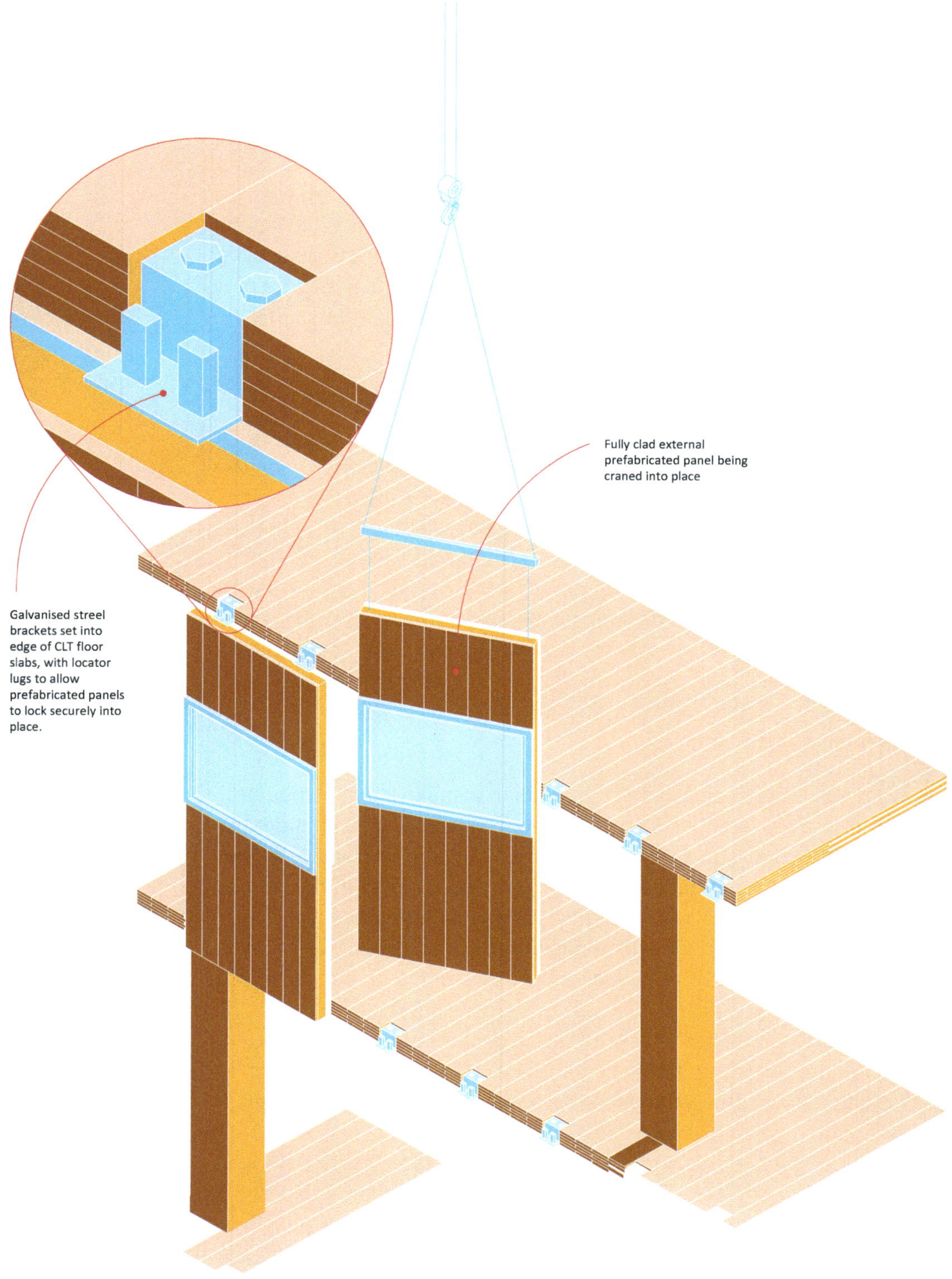

Figure 13.1 Prefabricated façade cladding assembly.
Source: Hayes/Marriage.

Prefabricated cladding

As noted in the chapter on prefabrication, wall elements can be completed in the factory and brought to the building site fully assembled – external walls as well as internal walls. The key to all this is to have a construction system planned out and fully tested, so that all that is needed on site is to simply clip the cladding panels into place on site. The primary advantages of the windows and their weatherproofing being completed at ground level, rather than attempting to resolve matters hanging off the building on a scaffold high above the ground, is accuracy as well as safety. Assuming that the panels themselves have been manufactured to a high degree of quality, the issue of waterproofing then switches from the panel to the horizontal and vertical joints between adjoining panels.

Another advantage is the prefabrication of the cladding panels themselves, with installation often proceeding soon after the main structure of the building. The building structures can be assembled in rapidly quick time. Some of the advantages of constructing in mass engineered timber are the use of accurate manufacture, the tight tolerance of the engineered joints and the prefabrication of the structural materials. Overall though, it is important to stress that one contractor should ideally be responsible for the overall façade in terms of design, installation, and waterproof guarantees: this is not an area you want to be arguing over who is to blame if anything does not perform.

Maintenance and access

It is easy enough when designing your building to make some computer renders that show everything shiny and new. But you need to do more than this: you need to think about how to keep the building look shiny and new, and how you are going to gain access to the façade to keep it that way. In order to be able to undertake that regular façade maintenance, you will need to have a method of regularly accessing the outside. You will need to design this system in at the same time as you design the façade itself.

There are some options for this: scaffolding is one system, but an expensive, one-off operation, as it costs a lot to erect the scaffolding and it blocks out the views for quite some time while the building is covered. Another option is to have a lowerable swinging stage that is lowered from the top of the building and via this stage, both cleaning and maintenance can be undertaken. But while this is probably acceptable to inhabitants of commercial office buildings, the top floor of any residential building is likely to be a highly expensive penthouse and the owners are not the sort of people who want maintenance workers climbing over their luxury apartment and lowering a swinging stage. So, not a viable option either.

In New Zealand, cleaning is often done by daring climbers on abseiling ropes, in preference to relying on someone else's stage or building maintenance units (BMU). In some other countries, abseiling may be very much frowned upon, and BMUs may be mandated.

But there is an easier option: design safe maintenance points into the façade itself. A sleek clean façade like those on high-rise modern architecture is unlikely to be what you would want on a residential apartment, and along with the recommended balconies and sunshades on a residential building, it may be easier to simply let the maintenance crew access the outside of the building directly from the deck of the balcony on each floor. The key thing they will need is a strong and safe reliable access point: effectively a large stainless-steel certified hooking point on the inside of every balcony. In some cities, such as Vancouver, it is more common for each floor of apartment towers to have a small projecting lip on the periphery of the floor, enabling the window cleaner to walk around the outside, complete with a series of points to clip a safety harness onto.

Cladding fixings

Cladding systems need to be sourced specifically for each situation: what works on one building may not be suitable for the next one. Sutherland wrote extensively on this in 2019, noting the myriad of concerns that need to be considered when specifying and installing façade cladding. Key points to note are to observe the principles of the ventilated cavity behind the final cladding, and that the truly weatherproof layer needs to be watertight behind this. This was noted in Tall (Marriage, 2019) where some simple rules for cladding details were called out: PROBLEM and CHANGED (refer Figure 13.2). To these, post-Grenfell, has added the importance of the vertical and horizontal cavity barriers, as well as access for maintenance and repair. Any system that goes on, must stay on – but it must also be able to come off again, with minimum of fuss.

The simplest system is a gravity lock clip system, where cladding panels are fixed into position and held there by reason of gravity and locked permanently into place by the next overlapping panel. This needs to resist both seismic action (i.e. upward thrust and sideways movement) as well as the ordinary (and extra-ordinary) wind movements expected. Terracotta cladding systems operate on this principle – the panels are simply hooked over the support brackets and locked into place by the next terracotta panel above. In this case, the terracotta façade will be acting as a rain-screen cladding only, and while it will rid the wall of 95% of rain, it will let some through the gaps. Hence the design and construction of the backing layer is crucially important in order to stop all moisture ingress at this layer. It is also the key area to let the façade breathe moisture out as well, so ensure flashing details are accurately installed around the window area (Figure 13.3).

P.R.O.B.L.E.M

Pressure :
create a chamber in your façade design to allow the pressure to equalise

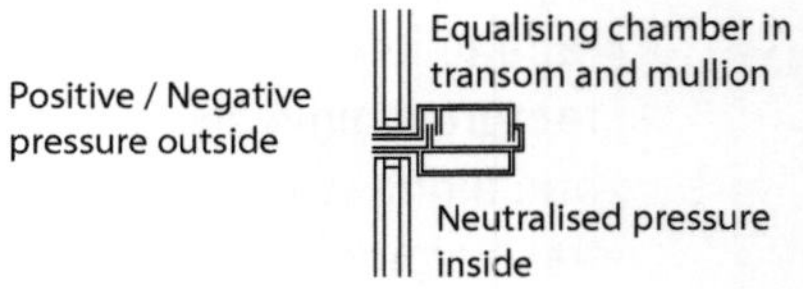

Rebates :
don't let water drive straight in, form a physical step in the slab and the walls

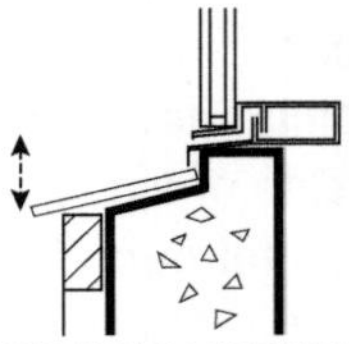

Overlap :
the lower member should always be protected by the upper member

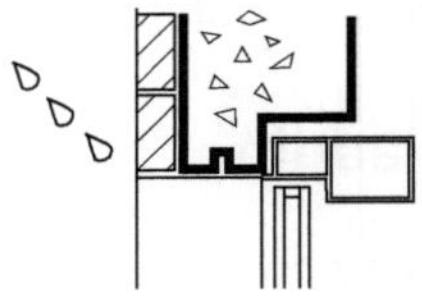

Baffle :
provide a second or third line of defence with an internal barrier

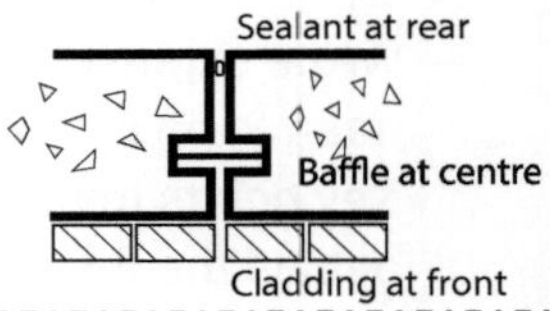

Lips :
rebate the slab edge to form a lip that protrudes out

Exit :
always allow the water to find a way out and exit the building façade

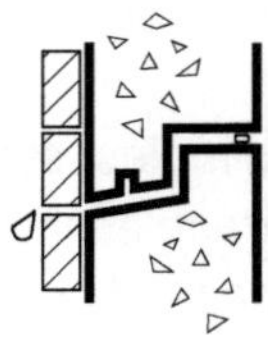

Membrane :
always lay a waterproof membrane (DPM or DPC) on your base material

Figure 13.2 PROBLEM.
Source: Marriage (*TALL*).

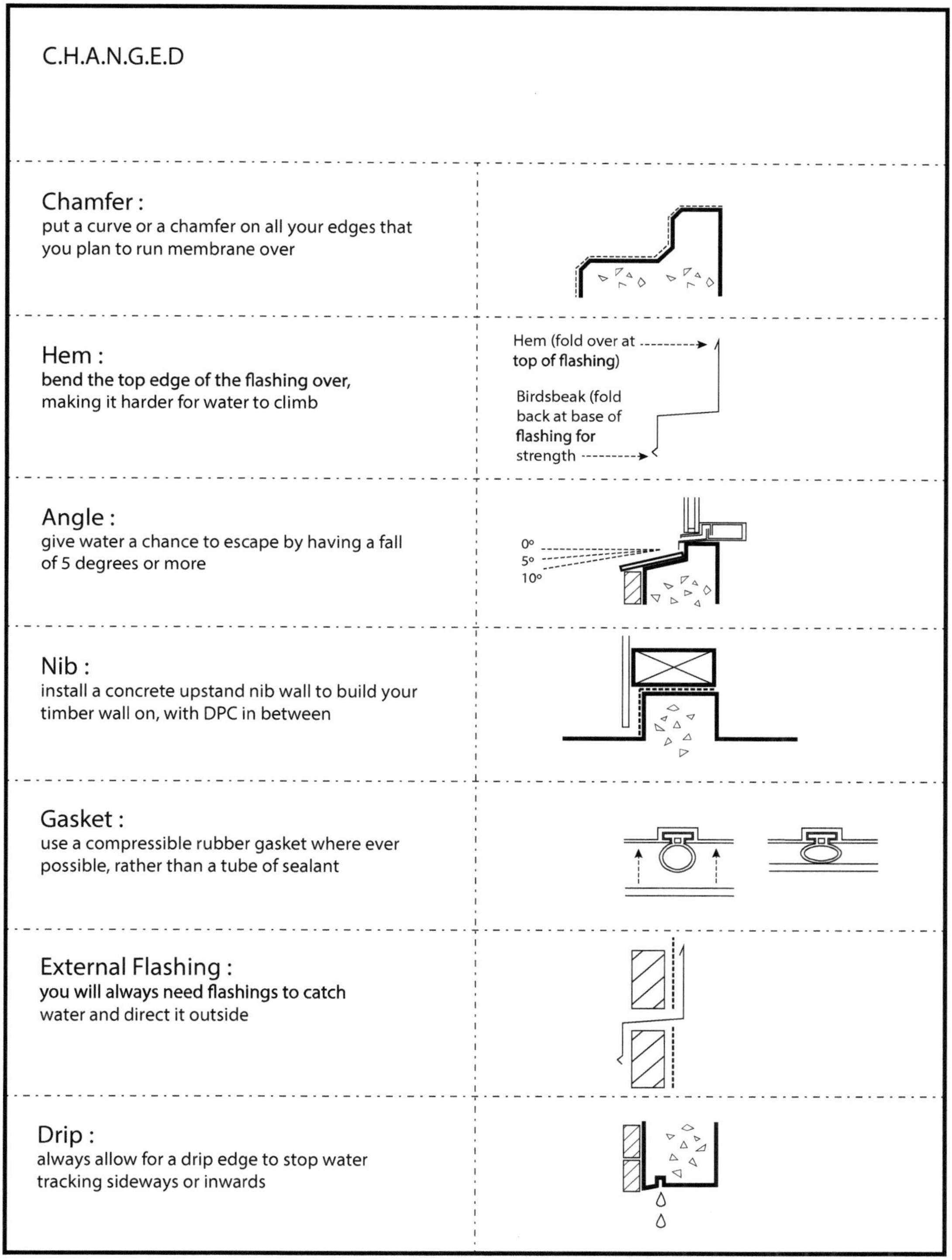

Figure 13.3 CHANGED.
Source: Marriage (*TALL*).

Window details

The success (or failure) of any building scheme will ultimately rely on it being able to successfully resist leaks. An apartment can definitely have too many windows, as the residents of the apartment building next to the Tate Modern discovered in London: people can look in as much as you can look out (Sturges, 2019). Apartment buildings tend not to be in smooth glass towers (except perhaps in Dubai due to the intense heat), apartment buildings are modelled to create depth, shade, and visual interest. As an antidote to an excess of glazing, external louvres may be used. This system is demonstrated with great effect on the *Apartamento il Casino* building in Wellington (architect: Studio Pacific, n.d.). Here, on the sunny northern façade, an array of carefully placed horizontal metal louvre blades give both shading and privacy to the apartments within. To avoid excessive heat entering the building on its sunny elevations, deep reveals and overhangs are frequently used on the façade, to provide protection to opening windows from wind and rain, and to help lessen the wind-blown effects of the height of the building.

Other penetrations

Residential building façades are far more complex and convoluted than the comparative simplicity of an office building. While an office façade may insist on a simple one-surface envelope, with few (if any) external access points like balconies, any residential façade needs to be a far more complex beast in order to provide fresh air, sun shading, wind shedding, privacy inwards, and views outward, as well as localised maintenance regimes. Again, simple guidance rules apply: allow a route for wind-driven rain to exit the building fabric and for the cavity to dry out. The Case Study chapters will show these some of these window installations in more detail. Key amongst these issues is the interface between the balcony and the wall, an area often neglected, to the building owners' regret. It is crucial to carefully design and install the flashings around penetrations like balcony brackets so that no water damage can occur here. The next chapter is all about the balcony.

Works cited

Barrett, Dave. 1998. The *Renewal of Trust in Residential Construction*. Canada: Commission of Inquiry into the Quality of Condominium Construction in British Columbia. Available at: http://www.qp.gov.bc.ca/condo/c3_iii.htm

Bonner, Matthew, Wegrzynski, Wojciech, Papis, Bartlomiej, and Rein, Guillermo. February 2020. "KRESNIK: A top-down, statistical approach to understand the fire performance of building facades using standard test data." In *Building and Environment*, Vol. 169, 106540. Elsevier. https://doi.org/10.1016/j.buildenv.2019.106540.

Available at: https://www.sciencedirect.com/science/article/abs/pii/S0360132319307528?via%3Dihub

Donn, Michael and Thomas, Grant. 2001. *Designing Comfortable Homes: Guidelines on the Use of Glass, Mass and Insulation for Energy Efficiency*. Wellington: EECA/CCANZ.

Dyer, Peter. 2019. *Rottenomics: The Story of New Zealand's Leaky Buildings* Disaster. New Zealand.

Grenfell Tower Inquiry, 2019. *Grenfell Tower Inquiry: Phase 1 Report*. The Rt Hon Sir Martin Moore-Bick. London: Her Majesty's Stationery Office.

Marriage, Guy. 2019. *Tall: The Design and Construction of High-Rise Architecture*. London: Routledge.

Murray, Jessica. 2020. "Manchester tower block residents ineligible for £1bn recladding fund." *The Guardian*, 31 May 2020. Available at: https://www.theguardian.com/uk-news/2020/may/31/manchester-tower-block-residents-ineligible-1bn-recladding-fund-skyline-central

Ramsdonk, Thomas van. 2020. *A New Zealand based Study on Airtightness and Moisture Management*. Wellington: Proclima. Available at: https://proclima.co.nz/pro-clima-study/

Studio Pacific. n.d. *Appartamento il Casino*. Available at: https://www.studiopacific.co.nz/projects/2013/appartamento-il-casino/

Sturges, Fiona. 2019. "Tate Modern's neighbours miss the point: cities make voyeurs of us all." *The Guardian*, 16 February 2019. Available at: https://www.theguardian.com/commentisfree/2019/feb/16/tate-modern-angry-neighbours-cities-voyeurs

Sutherland, John. 2019. "Façade: design." In *Tall: The Design and Construction of High-Rise Architecture*, edited by Marriage, Guy. London: Routledge.

Taylor, Diane. 2020. "Homes in London high-rises left 'wet and cold' after cladding removed." *The Guardian*, 22 January 2020. Available at: https://www.theguardian.com/society/2020/jan/22/homes-london-high-rises-left-wet-cold-cladding-removed

14

Chapter Fourteen

Balcony – the garden in the sky

Balcony – the garden in the sky

Guy Marriage

DOI: 10.4324/9781003123873-14

Introduction

Despite what the weather may be like in the city you are designing your apartment building for, access to some private space outside is a crucial requirement for inner-city apartment living. The very simplest form of this is the balcony, which can come in many sizes or design considerations. Well-designed, they can be an inexpensive addition to a building project that both brings joy to the inhabitants within and helps animate the façade outside (Poon, 2020). Poorly designed, they can be too small or otherwise inadequate, and if balconies are really badly designed they can be a source of leaks into the building fabric, threatening the entire structure of the building itself. What has become self-evident over the Covid worldwide lockdown is the vital importance of having a balcony at all (Lufkin, 2020). Access to the outdoors while cooped up inside, was a crucial factor for good mental health. Being locked up indoors without access to the outdoors in any meaningful way is simply cruel.

Balcony design

The challenge is to design and construct a balcony that is adequate for the use of the intended inhabitants according to the Goldilocks principle: neither too small, nor too big, but one that is just right. It needs to work as an extension to the living area within the apartment while also having a good outlook over the public streets below. It will be used as a breakfast bar by some people if it faces east, or an all-day lunch spot if it faces towards the sun, while it will definitely be used as a dinner and drinks spot if it faces west. A balcony sited on the cold face of the building, away from the sun, will likely never be used at all, so think carefully.

A balcony may be used by some simply as a place to grow pot-plants or vegetables, while others may just use it to store bicycles and miscellaneous items. For some people, it will be their pride and joy and the main reason they bought the apartment, spending hours out on the edge poised above the road, while for others it may be seldom used, except for smoking. Regardless of the usage, it needs to be designed well in the first place.

One of the primary decisions is whether to have balconies recessed into the bulk of the building, or alternatively to have them projecting out into the airspace above the footpath/sidewalk below. This decision has a large effect: recessed balconies can be a lot warmer and more comfortable, as the wind tends to move sideways across the face of the building to such an extent that even driving rain will not enter onto the balcony deck; while with an external balcony, the climate effects are usually much more noticeable: total immersion in the wind and rain (Figure 5.2). The design of recessed balconies simply implies a step back from the façade and potentially the addition of a nib on which the frames of the glazed windows and doors can seat (Figure 14.1). This system is often used when an

Figure 14.1 A range of several balcony types.
Source: Hayes/Marriage.

existing building is converted to residential. Balconies can be a secure private place while also being very public: make sure that burglars cannot encroach on this (Manaugh, 2016).

Balcony projection

A projecting balcony will likely get more sun and also a lot more wind and even snow, if that is a weather condition in your area. Projecting balconies require very careful detailing to avoid problems but can be achieved in various ways:

Firstly (and simply), the use of a rigid frame structure allows for cantilevering the beams a few meters beyond the façade with sufficient back span being provided. The inner slab then continues outside, simply seating on top of the beams and thus creating the balcony's area. With careful consideration of the drainage plane, this method can achieve great integration with the internal spaces.

The second option results from the interior floor slab projecting outside the façade without the need for cantilevering beams. This can be achieved with almost any flooring system as long as the perimeter beam provides sufficient support and the deflection is reasonable. The use of flat slabs and flat plates is also a relevant and easy alternative in achieving a balcony with sufficient cantilevering and free of beams. The key issue here is waterproofing: it is (obviously) vital to stop any water ingress on the slab at the junction with the external façade. Raised floors are a vital part of this detail, both inside and out.

The third option draws upon the design of bolt-on balconies where the balcony possesses its own structural components fixed to the building's primary structure. This is commonly designed as a series of closely spaced steel beams bolted to the slab or perimeter beam of the building and cantilevering from it. This type of balcony must be careful designed to limit thermal bridging and water penetration, but can work well, with the balcony simply craned up into place on completion of the project. Some balconies are bolted on simply through connections at floor slab level, while others utilise the structure of the whole balcony, taking strength from connections at a structural handrail level as well at slab level.

Balcony orientation

Some developers try to resist the need for balconies, noting that they can be expensive to build and may not appear to be occupied that frequently, but balconies have an integral part to play in the good design of a successful apartment. Omrani et al. (2017) note that the natural ventilation of a single-side-ventilated apartment can be improved by the provision of opening doors onto an open balcony. The orientation of the building to the prevailing wind and the size of the balcony can have significant effects on air circulation within the apartment, especially if rooms can open onto the outdoor space from more than one direction. Even if a dual-aspect apartment layout is not possible and only a single-aspect apartment can be achieved, then a well-designed balcony is still paramount to the comfortable living of the inhabitants within.

Clearly, orientation of the balcony towards the sun is important in colder climates, while a certain amount of shading is important in hotter environs. With stacked balconies, a large amount of self-shading will naturally occur from neighbouring balconies above; less so if they are staggered in layout. An exposed external balcony will be affected by wind far more than a recessed or semi-enclosed balcony, so consider whether or not you want to encourage more wind into the apartment.

Balcony construction

Firstly, the structure itself needs to be extremely strong. Balconies may be rarely used at times, but when there is something to see out the window, often everyone will rush out to see what the commotion is. The last thing you want is for the balcony to collapse: this has happened all too often, and often with catastrophic consequences (CBS News, 2003, 2015; Davidson, 2019; Jarosz, 2017; McNeilly, 2017). Large apartment buildings, of course, must be built to stronger standards than smaller residential house balconies as they are subject to higher wind loads, higher seismic acceleration, and potentially greater live loadings, as well as having potentially greater consequences if failure were to occur. Structural independence of each balcony is also crucial, so that a Ronan Point style disaster cannot happen: there must be no reliance for support on the balcony below. The best way of ensuring strength is to make the balcony an integral part of the building itself.

The balcony will need the input of a Structural Engineer (SE), as it has to resist considerable loads – both gravity loads (i.e. from many people standing on top) and lateral loads (from people pushing outwards on the handrail in crowded situations). The balcony acts as a cantilever, while the balustrade further utilises that cantilever principle to the maximum. Mistakes can be fatal.

There are only a few key parts of a balcony: the floor, the balustrade, the handrail, and the threshold back into the apartment (Figure 14.2).

Balcony floor

Floor slabs may be extended out to form the floor of the balcony, but utmost care must be taken to ensure that this does not create any paths

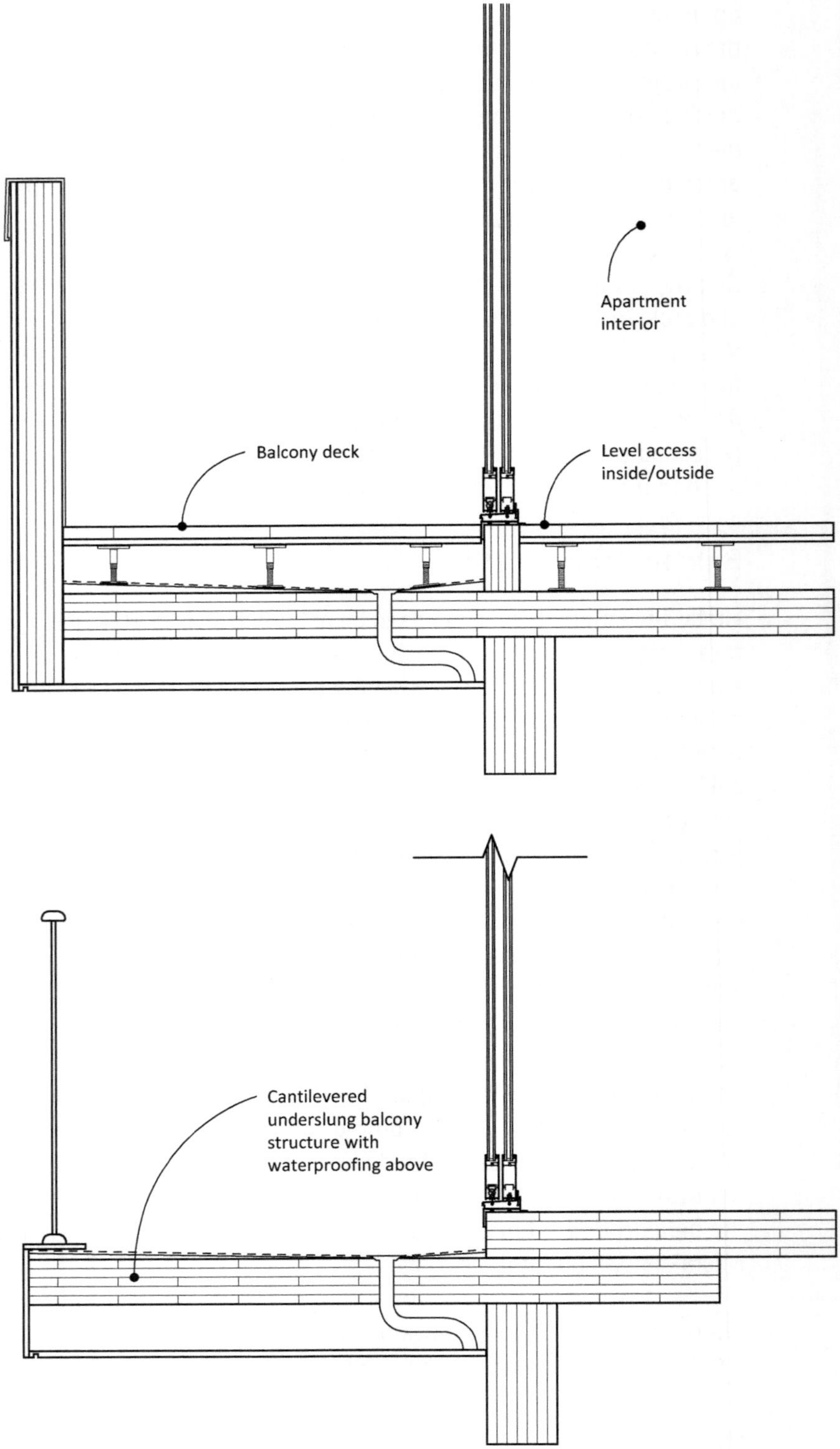

Figure 14.2 CLT balcony construction options.
Source: Hayes/Marriage.

for water ingress. Crucially for the floor design, it should capture any rainfall and conduct it carefully away without appearing obvious. Some designers make balcony floors from perforated metal so that the rain can wash through: sadly though, this also means the balconies below will also benefit from dust, dirt, spilled coffee, cigarette butts, etc. as well. My recommendation is: don't do that. Solid floors are best. The waterproofing must be bullet-proof in terms of detailing, ensuring no water can make its way inside the building envelope. Designed well, a balcony can have an almost seamless threshold to the living room within the apartment and a flat flush finish to the balcony itself, for ease of use and ease of maintenance. Refer to our case study chapters for some good examples. Think carefully about where rainwater goes at an early stage in design – drainage may need to be hidden within the balcony soffit, with downpipes descending within the façade nearby.

Balustrade

The balustrade is the highly visible part of any balcony design, rising vertically up from the floor of the balcony. Some architects prefer a fully glass balustrade, but this is seldom a very good idea. While balconies are part of the external space of an apartment, they also function to give privacy to the inhabitants within, both from people below on the street as well as other viewers from neighbouring buildings. A full height glass balustrade stops any possibility of privacy and therefore limits the eventual usefulness of the entire balcony. It also takes the brunt of the weather and can require frequent cleaning, both inside (easy to clean by the inhabitant) and outside (a real recipe for disaster if people are leaning outwards over the handrail). Cleaning implements and soapy water can fall to floors below, the glass can shatter, leaving a large dangerous gap, and in worst case scenarios, the apartment dweller themselves could topple to the ground. My recommendation: solid balustrades, at least to 800mm high (2′ 6″) so that someone seated inside can still see out to the horizon, with small rails or glazing above that to take it to the required minimum handrail height.

Balustrade height

What is the minimum balustrade handrail height? This will vary from jurisdiction to jurisdiction, but common sense also comes into play. While 1,000mm (3′ 3″) may satisfy the building code in theory, or whatever your local design/building code requires; on a really tall building, the minimum recommended height might be 1,200mm (3′ 11″) or even 1,400mm (4′ 7″). Provide a height that is appropriate to the risk. Wind uplift can be considerable on very tall buildings.

Baluster posts

If your design is not fully glazed but requires solid or semi-solid balustrading, baluster posts will be required at regular intervals to support the balustrading. A baluster post may be bolted onto the edge of the balcony; a simple principle is to never rely on just one bolted connection (a single connection may pivot around a point), so always have two rows of fixings. Make sure all fixings are through the side or underside of the balustrade, not down into the substrate from the top, as that can be a route for water damage to occur.

The balustrade materials need to be considered closely with the design itself, as some are more suitable than others. Horizontal balustrading is forbidden under some building codes, while allowed under others. Vertical balustrading is always acceptable, as it means there is less chance for tiny people to climb it. Flat sheet metal or perforated sheet metal may work well, reliant on satisfactory fixing design. Remember that any fixings in an exterior environment will be subject to accelerated corrosion, so think about ease of accessibility to fixings for maintenance and replacement.

Any gaps in the balustrade should ensure there is no chance for a baby to crawl through or fall through the gap. Official design standards will refer to a 100mm diameter sphere to equate to a baby's head size (a sphere of 100mm diameter has a circumference of 314mm, comparable to a baby's head size at birth of between 320 and 370 mm circumference). But by the time they get to crawling or climbing (from the age of about six to nine months), babies heads are far too large to make it through a 100mm gap, with a circumference of 400–445mm. This implies a gap of 130mm should be the minimum, not 100mm (WHO, n.d.). There is therefore literally NO chance that a child at crawling or walking age could fall through a balustrade by accident. However, toddlers and young children can climb just about anything, so design the balustrade very carefully to make sure that they cannot climb this. Provide no gripping points for tiny hands.

Handrails

Handrails complete the balustrade, sitting on top of the balustrading and taking the bulk of the horizontal and vertical forces. At the very least, if you have a glass balustrade, consider fixing a stainless-steel handrail to the top for visual and physical restraint. Think about the top surface of the handrail too: a slight slope inward is best, to stop anyone placing anything on the handrail that could fall outward. Again, ensure that any fixings for the handrail are from the side or from below, rather than from the top, to lessen any chance of water gaining inroads into the balustrade below.

Ensure that the handrail material is as long-lasting as possible, i.e. use bead-blasted stainless steel, thick section powder-coated aluminium, or a long-lasting profiled hardwood in preference to mild steel that will need continued maintenance.

Threshold

A flush threshold onto the balcony is most desirable but is only achievable with a well-considered weatherproofing design. If there is an extensive overhang above the threshold (i.e. like 2m of balcony overhead) then a small 20mm lip may suffice and still work as a fully disabled-accessible junction. If there is any chance of wind and rain pushing water across the balcony and up to the door threshold, then you must create a gap of at least 10mm horizontally to a drain below, or incorporate an offset in height. This is easily achieved by having a grill across the door front – but this requires the main structure of the balcony to be lower by 100–150mm. Refer to the details shown in the case study chapters for a variety of solutions.

Balcony installation

The quickest manner of balcony installation is to fully fabricate the balconies in a factory and crane them into position late into the completion phase of the project, requiring rigorous planning and careful forethought. Each balcony will need to have its own structural solution and should not rely on other balconies above or below. On one of our projects, we attached 25 balconies to the exterior face of a completed building in a single night to avoid extensive road closures to the main transit route below – only meticulous planning and preparation saved us from clogging the streets below the next day. So, plan the structural connection system rigorously – get the details right and it works well.

Balcony storage

Intentionally or not, balconies become storage areas for many apartment residents: it is the outside, left-over space that doubles as a 'back yard', yet the effect on the front of a building is often far more 'front yard' than intended. Consequently, while some residents may stage their balcony to present a polished view to the world outside, others tend to accumulate disused broken equipment and display that to the world instead. Bicycles are a classic occupant of many balconies, with bicycles being a difficult item to store indoors in a small apartment. Some residents try to build in their balcony with glazing (a popular move in Hong Kong) to gain extra 'inside' space, but this is usually discouraged or disallowed by the building owners.

In any event, our recommendation is to have solid walls to the balconies to avoid any view through and to cover up the human-generated detritus of modern apartment living. A protruding solid balcony also helps provide visual and aural privacy to the apartment residents within.

Lanai – indoor/outdoor space

You may not have heard of this term before but it is a Polynesian word, originally from Hawaii. It means an indoor/outdoor space – a porch with a veranda roof, used as a living room. While a veranda implies something long and relatively thin, a lanai is more of a room in itself and is a crucial design feature where the air is really hot and humid and not too windy. As an outdoor living room, it usually has no windows and hence it needs to be water-proof at floor level. You're not going to want one of these if you live in Scotland or Newfoundland, but they come in very handy in moist climates like Hong Kong, Auckland, Miami, and Sydney.

Roof gardens

A garden high in the sky can be the most wonderful place for an outdoor space, and for the lucky few at the top of the building, individual roof decks are possible. It is even better if all the residents can access the roof space, although this is so often not achieved. Communal facilities like roof gardens need loving care and attention to ensure that the plants are well watered and cared for, while an equal amount of care to the waterproofing layer should ensure the building's structural integrity is maintained. Regardless if it is private or public, any roof garden must ensure good waterproofing over the top of excellent structural integrity. A concrete slab is best rather than a lightweight build, to enable a weathertight rooftop. Common access up onto the rooftop must be designed with as much care as is possible to make sure this is not a source of possible water ingress.

The weight of planters and the vegetation within them can be significant. Waterlogged soil can entrap significant extra weight which, being at the very top of the building, is a seismic concern (one cubic metre of water-logged soil can weigh more than one tonne). As an architect on the project, make sure that waterproofing systems are fail-safe and cannot let water travel inside the building. Ensure that any earth for trees and plants is based on a lightweight pumice-based soil substrate, rather than heavy clays and rocks. Keep the thickness of vegetation planters to a minimum to help the plants survive and remember that plant roots can work their way into the smallest cracks and cause havoc. Use a seamless plastic liner that cannot be breached and plan for a back-up waterproofing solution in the event that the roof-garden breaches your 'unbreachable' liner.

Works cited

CBS News. 2003. "Without warning, chaos. Who's to blame for a Chicago Porch collapse that killed 13 people?" *CBS News*, 29 June 2003. Chicago: 13 people killed, 57 injured.https://www.cbsnews.com/news/without-warning-chaos/

CBS News. 2015. "Deadly balcony collapse tied to rotted wooden beams." *10TV WBNS*, 17 June 2015. California: 6 people dead.https://www.10tv.com/article/deadly-balcony-collapse-tied-rotted-wooden-beams

Davidson, Helen. 2019. "South Korea: world championship athletes injured in fatal balcony collapse." *The Guardian*, 27 July 2019. South Korea: 2 people dead, 16 injured.https://www.theguardian.com/world/2019/jul/27/south-korea-balcony-collapse-deaths-world-championship-swimmers-injured-

Jarosz, Brooks. 2017. "Condo balcony collapse prompts 150+ decks deemed 'unsafe'." ABC 6, 10 June 2017. Hayden Run, Colorado: 2 people injured.https://abc6onyourside.com/news/local/condo-balcony-collapse-prompts-150-decks-deemed-unsafe-06-10-2017

Lufkin, Brian. 2020. "What outdoor space tells us about inequality." BBC WorkLife, 18 June 2020. Available at: https://www.bbc.com/worklife/article/20200617-the-luxury-and-privilege-of-a-balcony-or-yard-during-covid

Manaugh, Geoff. 2016. *A Burglar's Guide to the City*. New York: Farrar, Strauss and Giroux.

McNeilly, Hamish. 2017. "Multiple failings revealed in Dunedin Six60 gig balcony collapse." *Stuff*, 3 February 2017. Dunedin, New Zealand: 18 injured, 5 seriously, with one person having severe spinal injuries.https://www.stuff.co.nz/national/89073303/multiple-failings-revealed-in-dunedin-six60-gig-balcony-collapse

Omrani, Sara, Garcia-Hansen, Veronica, Drogemuller, Robin, and Capra, Bianca. 2017. "On the effect of provision of balconies on natural ventilation and thermal comfort in high-rise residential buildings." *Building and Environment*, 123, 504–516. https://doi.org/10.1016/j.buildenv.2017.07.016

Poon, Laura. 2020. "A lesson from social distancing: build better balconies." *Bloomberg CityLab*, 21 April 2020. Available at: https://www.bloomberg.com/news/articles/2020-04-20/lesson-from-coronavirus-build-better-balconies

World Health Organisation. n.d. "Head circumference-for-age (birth to 5 years)." WHO Child Growth Standards.

15

Chapter Fifteen

Exporting apartments – lessons in prefab from the north

Exporting apartments – lessons in prefab from the north

Joseph Sturm

DOI: 10.4324/9781003123873-15

Introduction

Situated on the Baltic sea and variably described as 'eastern' or 'northern' European, Estonia has seen rapid development since restoration of independence in 1991 and continues to position itself as an innovative, digitised and business-friendly modern European state. Combining its extensive timber resources, modern technology and an export-oriented economy, Estonia has emerged as the European Union's largest exporter of prefabricated timber buildings, with nearly 90% exported in 2019. The emergence and continuing advancement of the sector is a significant example of the country's successful post-Soviet development. From beginnings as a source of low-cost labour coupled with deregulation in the Nordics and onward to development of a strong brand with a shift towards skills, quality and added-value. This chapter will document the background, current situation and potential of the Estonian prefabricated timber building sector before discussing their role in construction of apartment buildings and lessons learned able to be applied elsewhere.

Estonia has a long history of timber housing with log houses being a predominant construction method until the early 20th century (Metslang, 2016) before transitioning to stud-wall construction (Riistop, 2016). Most notable among these, are the *Lendari*- and *Tallinn*-type houses. These two and three storey apartment buildings with 8–12 units were developed in the 1920s and 1930s to provide workers' accommodation in the years following independence from the Russian Empire when Tallinn, the Estonian capital, saw rapid growth (Vaitmaa, 2016).

The Second World War brought Soviet occupation and during this time timber construction declined in favour of concrete and brick (Metslang, 2016). War damage as well as the arrival of migrant workers from across the Soviet Union created a housing shortage and from the 1960s to the 1980s many standardised prefabricated concrete panel houses typical of Soviet cities were built (Hess and Tammaru, 2019). Looking beyond residential construction, a small research institute for developing timber structures remained at Tallinn Technical University and a number of timber shell constructions were built (Ehasalu, 2015; Metslang, 2016). Additionally, a laminated timber factory was established in Põlva in 1979 to produce timber structures for sports arenas, industrial buildings and bridges (Metslang, 2016).

Beginnings of timber prefabrication

Following dissolution of the Soviet Union and with it, rigid centralised economic planning, Estonia quickly transitioned to a market-based economy. The 1990s saw rapid economic growth, stimulated by liberal business-friendly policies such as flat tax and no corporate income tax on profits reinvested back into the same company. Combined with a favourable location close to Finland and Sweden as well as the decision to re-establish the Estonian Kroon pegged to the Deutschmark, the country quickly attracted foreign investment achieving the highest growth rates amongst former-Soviet countries (Laar, 2007; Lumiste et al., 2008).

In the construction sector the role of timber once again began to grow, and introduction of free enterprise led to the formation of many new companies (Riistop, 2016). Given the small size of the domestic market, many new enterprises quickly reached the conclusion, that to grow, they would need to export (Mitt, 2016). During the same period, many Estonians left to seek work abroad, attracted by higher salaries in western and northern European countries (Tammur et al., 2017). Conversely western and northern European employers were attracted by lower wages in Estonia and took the opportunity to outsource production there with Finland and Sweden representing Estonia and Finland's largest source of foreign direct investment (Lumiste et al., 2008).

Meanwhile, in the Nordic countries, the 1990s saw a period of deregulation in the construction sector. Previously prescriptive building codes rigidly enforced a maximum two storeys for timber buildings in Finland and Sweden, and three storeys in Norway. Starting in Sweden in 1994, each deregulated with a shift to performance-based building codes. This meant the number of storeys was no longer restricted, but rather conformance with specific technical requirements had to be demonstrated (Kallas, 2018).

By the mid- to late 1990s, this combination of a changed economic and regulatory climate had generated the factors necessary for Estonian companies to begin exporting prefabricated buildings to the Nordic countries: the Estonian construction sector had developed new enterprises with competence in timber buildings; there was increased investment, trade and business relationships between Estonian and the Nordic countries; and regulatory changes in the Nordics meant the demand for housing using timber structures was growing.

Organisation, systemisation and development of competitive advantages

By 1999, seeking to develop the potential for export of prefabricated timber buildings, 17 companies formed the *Eesti Puitmajaliit* (Estonian Woodhouse Association). The association's founding members had the aim to enable collaboration between Estonian timber building producers, to train staff, develop the sector's reputation, and in particular, grow export volumes (Estonian Woodhouse Association, 2019; Kallas, 2018). Building on the successes of the 1990s, in the early 2000s, the sector continued to grow. Development continued to be focused on the Nordic countries and this is reflected in the breakdown of export destinations, where these countries and Germany making up a significant majority of exports, Norway remaining the largest market with 31% of all exports by 2017 (Estonian Woodhouse Association, 2019).

Research seeking the reasons for the high proportion of exports to Norway, Sweden and Finland show many factors. The most obvious and often mentioned are price and geography. Interviews with owners and managers of Estonian timber prefab producers showed that Estonian producers' production costs allowed them to compete on price, while Estonia's closeness to these markets meant that transport is cost-effective and convenient (Kallas, 2018; Mitt, 2016). Moreover, producers' entry into these markets was helped through using personal connections and building up a local presence – mostly via distributors, agents and partnerships with locals.

Price and geographical closeness, however, are not the only factors, and the state of the market and government intervention have also helped create favourable conditions. Producers noted that entry to Nordic markets was simplified by the long history of timber construction in these countries. Clients are already familiar with timber and its benefits, which contrasts to the often sceptical attitudes in central and western Europe as well as their home market. State programmes further pushed the use of timber as a construction material. For example, in 1997 Sweden launched a wood, construction and furniture programme and through the 2000s the state developed many more timber-related programmes. The Swedish Wood Council, an organisation aiming to spread knowledge about and encourage use of timber, was established. Many universities developed doctoral programmes focusing on timber construction and actively collaborated with the industry. Similarly, Finland has propagated timber construction since the 1990s and government intervention significantly raised the proportion of timber-framed residential construction. Government procurement was used to create example projects using timber and as in Sweden, changes to university curricula increased knowledge of timber construction. Norway also followed a similar approach, and in 2008 the Norwegian Wood Programme was established with the aim of gathering all sectors, local government, business and research institutes to develop multi-storey timber construction (Kallas, 2018; Mitt, 2016).

In addition to favourable geographic, economic and regulatory conditions, the efforts of Estonian producers have also been a significant factor. Economic growth and with it an increase in wages has meant that Estonian producers can no longer compete on price alone. The Estonian market faces competition from Eastern European neighbours, who have developed their own timber prefab capabilities and are able to undercut on price. Estonian producers, however, are regarded as being able to deliver higher quality and have built a strong reputation. Moreover, many Estonian producers are willing to undertake technically complex projects as well as those requiring bespoke solutions (Mitt, 2016).

Current status

Today Estonia's timber prefab sector has matured into a well-developed, reputable and technologically advanced industry with around 170 companies active on the market. According to data by Eurostat, as of 2019 annual exports stood at around 427 million Euros which made nearly 90% of revenue (Estonian Woodhouse Association, 2020). With the exception of a blip during the global financial crisis of 2008, growth has been consistently strong. As well as the Nordic markets, growth has been further strengthened in Germany, Holland, Italy and Czechia (Estonian Woodhouse Association, 2019).

Producers offer a wide spectrum of products, which have been categorised in six typologies by the Woodhouse Association, three of which are applicable to multi-unit housing. The simplest of these are the *timber-framed* category referring to prefabricated stud wall constructions. Secondly the *element house* category features wall (and in some cases, floor and roof) elements ready to be transported to site and assembled. Finally, the *room element* or modules category features entire room modules. These vary in level of prefabrication – with timber-framed having the lowest level requiring cladding, doors, windows and wall linings to be installed on-site. The element category includes doors, windows and cladding, while the modular category offers the highest level of finishing. The latter two may use simple stud wall construction, engineered wood such as LVL and glulam or solid timber construction such as CLT. Together, these categories make up 70% of prefabricated timber structures produced in Estonia (Kallas, 2018). The remaining categories are smaller in scale – the 'Garden house' category incorporates small non-habitable structures such as saunas. The 'machined log house' and 'handmade log house' categories are typically used only for standalone houses. Their significance, however, should not be underestimated, with many of today's larger producers having begun with these typologies (Estonian Woodhouse Association, 2019).

Most companies have, over time, consolidated to focus on particular typologies and construction techniques. Similarly, companies send the majority of exports to just one or two markets. The focus on particular construction types has been driven by the search for operational efficiency through standardisation while the focus on specific markets is attributed to the importance of reference projects, referrals and personal contacts (Estonian Woodhouse Association, 2019; Kallas, 2018; Mitt, 2016). Additionally, differing rules, regulations and construction traditions in each country present a significant barrier to entry. This focus can often be seen through a company's name, such those showing Estonian-Nordic connections like *EstHus, EstNor* or Finnlogor referring to their primary means of construction such as *7 Element, Arcwood*, and *Timbeco*. While most companies active on the market are micro enterprises (defined respectively by European Union directive 800/2008/EU as up to 10 employees or 2 million EUR in annual turnover) or small enterprises (up to 50 employees or 10 million EUR turnover), a number of bigger players capable of producing multi-unit housing developments have emerged. The Woodhouse Association's 2019 selection of 33 case studies of "Leading Timber Construction Enterprises" gives 22 examples of companies with over 2 million EUR in revenue – 14 of whom show reference projects with multi-unit terraced or apartment typologies on their websites.

The range of apartment development projects evidences both Estonian producers' capacity to manufacture and export at volume as well as their reputation for taking on complex challenges. One Estonian producer, for example, supplied and installed timber apartment modules for the 14-storey building *Treet* in Bergen, Norway – the world's tallest timber building at the time of completion in 2015. Modules stacked up to four storeys high were installed within a glulam loadbearing frame (Abrahamsen, 2015). Additionally, the same producer claims to have built Norway's first (modern) four-storey timber-framed apartment building in 1999 as well as Norway and Europe's first five-storey timber-framed apartment building in 2003 (Kodumaja AS, 2019). In another example, an Estonian producer secured a contract to produce one of Finland's largest ever modular housing developments in Riihimäki, with over 300 apartments in nine four-storey buildings completed in 2020 (Astel Modular, 2019). Using the element typology, in 2016 the company Matek completed 127 apartments in buildings of five to seven storeys in the project *Slottet* on a challenging hillside site in Norway (Matek, 2021). Finally, the same producer supplied façade elements – another market segment growing in significance – as part of a pilot project to retrofit Soviet era buildings to achieve near Zero Energy Building (nZEB) status (Kuusk et al., 2019).

Looking to the future, Estonian producers are positioning themselves as a reasonably priced (though not the cheapest) option, while being able to deliver a high-quality product based on long term experience and competence. Many producers refer to rising costs of labour and materials as well have increased competition from neighbouring countries as significant challenges (Estonian Woodhouse Association, 2019; Mitt, 2016). To address these challenges, a need for increased automation as well as further training will be needed in the coming years. Increased environmental awareness and stricter energy efficiency standards are regarded as opportunities for the sector. To this end, in 2019 the Estonian government supported initiatives such as *PuitAIT* which seeks to increase collaboration between architects, engineers and wood house producers with the IT sector to digitise supply chains, marketing and the design process. Similarly, the Estonian Association of Architects received funding for *PuuExpo* to showcase Estonian timber design and increase collaboration with designers abroad (Enterprise Estonia, 2019). Addressing, producers' oft-stated frustration with a lack of acceptability for timber buildings on the domestic market, a breakthrough was achieved with a timber-framed building selected as the winning competition entry for the new headquarters of the Estonian Ministry for the Environment (ERR, 2019).

Discussion and conclusion

At first glance, Estonia's success in exporting prefabricated timber buildings appears to be a simple case of cheap labour and geographical convenience. While these factors along with Estonia's regulatory environment have undeniably been contributing factors and helped Estonian producers gain a foothold in new markets, this has been shown to be far from the full story. The sector has matured to offer diverse typologies and construction techniques, become technically skilled and able to deliver quality products. From this many lessons can be learned, and Estonia used as a case study for developing the timber prefabrication sector for apartment buildings.

Firstly, the effect of the industry's collaboration via the Woodhouse Association and Woodhouse Cluster created many benefits that no one company would have been able to realise alone. For example, resources were pooled to take part in trade fairs with high entrance costs, standard construction details developed and local regulations from Nordic countries translated. This mean that each member had access to resources otherwise only available to much larger companies while being able to focus their efforts on design, development, manufacturing and installation.

Secondly, Estonian producers' experience in Norway, Sweden and Finland show that even in markets long accustomed to residential timber buildings benefit significantly from government intervention to encourage using prefabricated timber structures for apartment buildings. Each country invested in timber construction programmes, committed to educating future construction industry professionals in the benefits of timber and led the way through public procurements to create pilot projects. For Estonian producers this contrast is particularly evidenced in their experiences – or lack thereof – with clients in their home market where timber is viewed sceptically when compared to concrete and steel and where government support for timber construction has, until recently, been minimal.

Thirdly, Estonian producers have not remained fixated on a race to the bottom and are not trying to compete on price alone. Producers have managed to set themselves apart by offering high quality, being flexible in offering unique solutions and willing to take on technical challenges. This has been underpinned by their commitments to their respective production methods and target markets, through developing their own specialist niche and building client relationships.

Together, the lessons from Northern Europe – be it the technical competence of the Estonian prefab sector or proactive measures by Nordic governments to support timber construction – show that where producers are willing to focus on quality, innovation and creating strong relationships, where industry associations are able to work for their members' collective benefit and where governments show leadership to encourage timber construction through pilot programmes, education and training; ever more high quality timber prefabricated apartments will be built, exported and erected.

Works cited

Abrahamsen, Rune B. 2015. "First 14-storey wood building in the world at Bergen in Norway." *5th Formu International Bois Construction FBC.* Nancy: Forum Bois Construction France.

Astel Modular. 2019. "Astel alustab Soome ühe suurima moodulitest korterelamuprojekti tootmisega (Astel to begin production for one of Finland's biggest modular apartment building projects)." *Astel Modular.* Accessed 27 March 2021. Available at: https://astelmodular.ee/astel-alustab-soome-uhe-suurima-moodulitest-korterelamuprojekti-tootmisega/

Ehasalu, Ave. 2015. "Rahvusvahelise turundusstrateegia arendamine Eesti Puitmajaklastri näited (Developing export marketing strategy: case of estonian wooden houses cluster)." Master's thesis. Tartu: Tartu University.

Enterprise Estonia. 2019. *Enterprise Estonia to Support Ten Projects of Professional Associations. 12 March.* Accessed 28 March 2021. Available at: https://www.eas.ee/enterprise-estonia-to-support-ten-projects-of-professional-associations/?lang=en

ERR. 2019. *Keskkonnamaja võidutöö näitab puidu kasutust parimal moel (Environment House Winning Entry Shows Timber Used in the Best Possible Way).* 20 March. Accessed 28 March 2021. Available at: https://kultuur.err.ee/921787/keskkonnamaja-voidutoo-naitab-puidu-kasutust-parimal-moel

Estonian Woodhouse Association. 2020. *Koroonaeelne aasta oli puitmajatootjate ekspordile edukas (The Year before Corona Was Successful for Timber House Producers).* 24 August. Accessed 27 March 2021. Available at: https://puitmajaliit.ee/uudised/koroonaeelne-aasta-oli-puitmajatootjate-ekspordile-edukas

Estonian Woodhouse Association. 2019. Kivil, Lauri. "Introduction" in *The Cutting Edge: Estonia's Leading Timber Construction Enterprises,* edited by Veronika Valk-Siska, 4. Tallinn: Estonian Woodhouse Association.

Hess, Daniel Baldwin and Tiit Tammaru. 2019. "Modernist housing estates in the baltic countries: formation, current challenges and future prospects." In *Housing Estates in the Baltic Countries: The Legacy of Central Planning in Estonia, Latvia and Lithuania*, edited by Daniel Baldwin Hess and Tiit Tammaru, 3–30. Springer.

Kallas, Jaanus. 2018. "Innovatsioon ja selle soodustamise viisid Eesti puitmajatootjate näitel (Innovation and the ways to stimulate it at the example of estonian wood house producers)." Master's thesis. Tartu: Tartu University.

Kodumaja AS. 2019. "Kodumaja Presentatsioon/Kodumaja presentation." *Kodumaja*, 27 May. Accessed 27 March 2021. Available at: http://www.kodumaja.ee/et/Presentatsioonid

Kuusk, Kalle, Phelo, Peep, and Kalamees, Targo. 2019. "Renovation of apartment buildings with prefabricated modular panels." *REHVA 13th HVAC World Congress CLIMA 2019*. Bucharest: E3S Web of Conferences 111.

Laar, Mart. 2007. "The Estonian economic miracle." *Backgrounder* (The Heritage Foundation), (2060), 1–12.

Lumiste, Rünno, Pefferly, Robert, and Purju, Alari. 2008. *Estonia's Economic Development: Trends, Practices and Sources*. Case Study. Washington, DC: The International Bank for Reconstruction and Development/The World Bank.

Matek. 2021. *Slottet*. Accessed 27 March 2021. Available at: https://www.matek.ee/et/project/slottet/.

Metslang, Joosep. 2016. "Vanemad palktarandid/Development of wooden constructions." rmt: *Puit Eesti arhitektuuris/Wood in Estonian Architecture*, Leele Välja, 17–42. Tallinn: Metsa- ja Puidutööstuse Liit.

Mitt, Arlyn. 2016. "Eesti puitmajatootjate tarneahelad välisturule sisenimisel (Estonian wooden house manufacturers supply chains for entering foreign market)." Master's thesis. Tallinn: Tallinn Technical University.

Riistop, Märt. 2016. "Uued puitkonstruktsioonid/New wooden constructions." rmt: *Puit Eesti arhitektuuris/Wood in Estonian Architecture*, Leele Välja, 43–54. Tallinn: Metsa- ja Puidutööstuse Liit.

Tammur, Alis, Tammaru, Tiit, and Puur, Allan. 2017. "Is there a migration turnaround taking place in Estonia? Migration trends 2000–2015." In *Estonian Human Development Report 2016/2017 Estonia at the Age of Migration*, edited by Raul Eamets, Kristina Kallas, and Tiit Tammaru, 9–10. Tallinn: Foundation Estonian Cooperation Assembly.

Vaitmaa, Ester. 2016. "Tallinna Maja." *Forte*, 30 September. Available at: https://longread.delfi.ee/forte/tallinna-maja?id=75652667

16

Chapter Sixteen

Case study 1: Clearwater Quays

Case study 1: Clearwater Quays

Pacific Environments NZ: Roger French and Enovate: Michael Newcombe

DOI: 10.4324/9781003123873-16

Figure 16.1 Clearwater Quays.
Source: Courtesy of Pacific Environments NZ.

Project title – Clearwater Quays Apartments, Rotoiti Lane, Christchurch, New Zealand (www.cqa.co.nz) (Figure 16.1)

Team info

Architects (and team): Pacific Environments NZ – Roger French, Phil Howard, Graham Allen, Mathew Philip, Brent Anderson, Sara Romana, Sarah Eising

Engineers (and team): Enovate Consultants – Michael Newcombe and Allister Moo

Truss designer: Buildable Layouts – Matt Jewson

Structural peer review: Engco – Bjorn Stankowitz

Mechanical: E^3 Building Workshop – Anton Smith

Electrical: E^3 Building Workshop – Mark Kessner, Simon Galea, Reuben Wood

Hydraulic: Conor Fullbrook, Ella Osborne

Fire: Olsson Fire & Risk – Per Olsson, Eva Andersson, Peter Wright, Tony Parkes, BECA – Fadi Jirjees, Jason Zhao

Façade designers/engineers: Inhabit Group – Gemma Craig, Christian Veloria

Acoustic: Marshall Day – Keith Ballagh, Shaun King, Tonkin & Taylor – Darran Humpheson

Civil: Engco – Andy Lind

Geotechnical: KGA-Robert Kamuhangire, Abilio Nogueira, Engco-Ragulan Kanapathappillai

Planners: Novo Group – Emily McDonald, Tim Walsh, Jeremy Phillips

Vertical transportation: Schindler – Robin Manderson

Contractors (and team): Construction Solutions (contracted by Logic Group) – Phil Tompkins

Project Managers and Q.S: Logic Group – Sam Cadden, Andrew Voss, Barry Lynch (QS), Eoin McLoughlin, Will Roberts, Julie Perkins, James Crawford

Developer – Clearwater Quays Apartments, part of Red Stag Group: Marty Verry

CLT supply – Xlam (Australia)

LVL supply – Timberlab

Prefabricated wall supply – Concision

Structural timber – Red Stag Timber

Exterior and interior cladding/finishing timber – Abodo

Date of construction – 2020–2021

Project info:

Number of storeys – two buildings, each with five floors of residential. No basement

Site coverage – Gross site area 3,740m^2 and gross floor area 1,700 m^2

Number of apartments – 10 x one/two-bedroom luxury apartments

Average size of apartments

Large apartment: 189.2m^2 for ground floor excluding decks and garages, including guest bedroom, 21.7m^2 porch

Large apartments for levels 2–5: 159m^2 excluding decks and garages, 20.5m^2 deck

Small apartment: 107.6m^2 for ground floor excluding decks and garages, including guest bedroom, 13.1m^2 porch

Small apartments for levels 2–5: 107.6m^2 excluding decks and garages, 18m^2 deck

Garage areas vary for each apartment

Approx. construction cost all up – $13.7m, of which $10.5m build cost (Figure 16.2)

Concept

Situated on the shore of the man-made Clearwater Lake, the Clearwater Quays residential apartments are orientated north to take full advantage of the mountain view and solar gain. The northern orientation also creates a connection from the apartments to the lake, via a communal jetty, enhancing the notion of resort-style living. When completed, each residential tower will offer a mix of 10 one- and two-bedroom apartments. The two residential towers are part of a staged construction, with the learnings from the first informing refinements for the second.

The project was commissioned by a company with an interest in the timber industry; naturally showcasing timber both structurally and visually became a driving factor of the design brief. It is ironic that the timber in many mass timber buildings is intangible, hidden by cladding and carpets, vital components in a building's acoustic performance. Our challenge was to seek out the opportunity to highlight the aesthetic and structural quality of timber. This was achieved via the dramatic, sculptural staircase

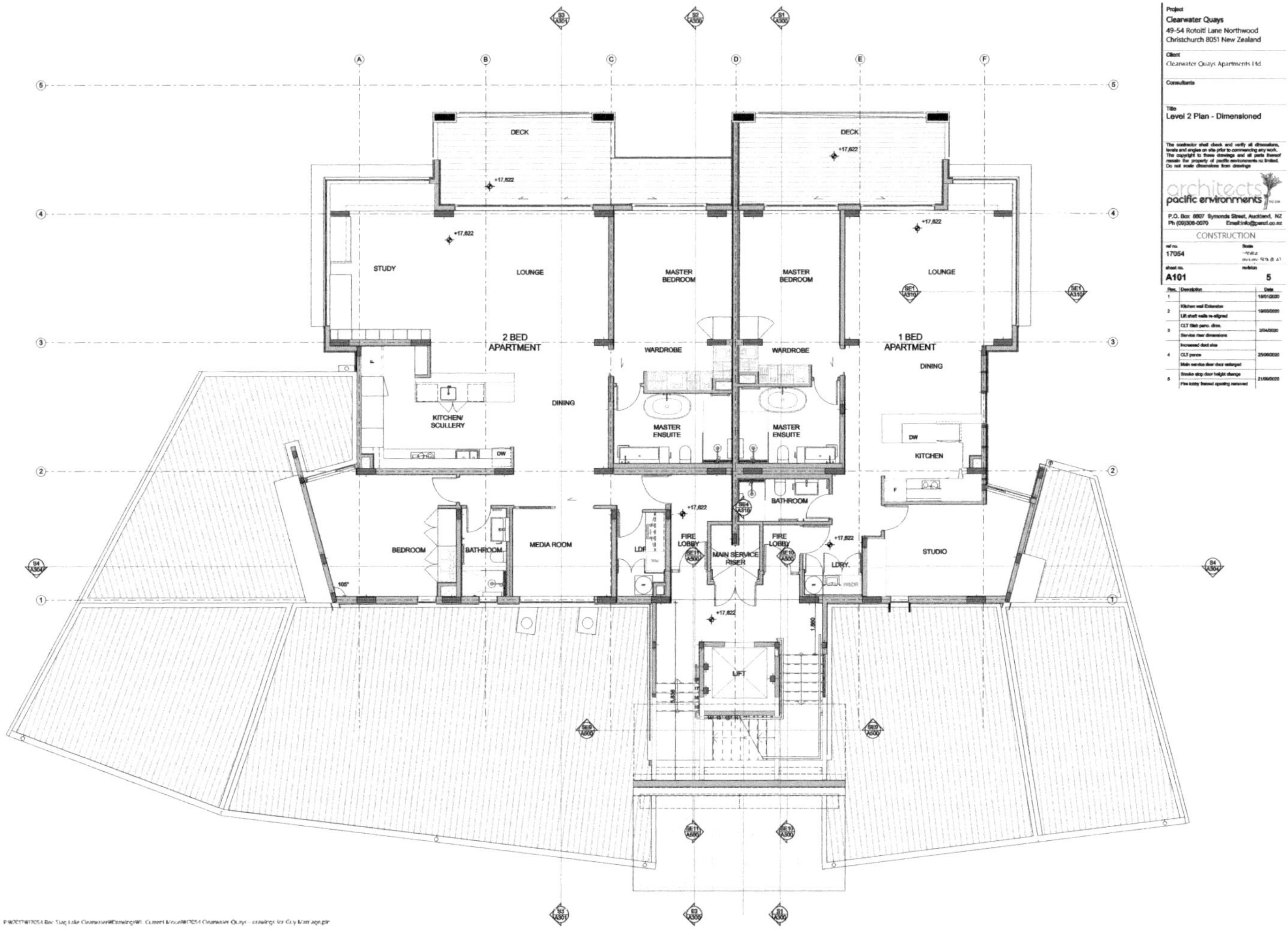

Figure 16.2 Clearwater Quays level 2 plan.
Source: Courtesy of Pacific Environments NZ.

structures projecting from the southern façade of each residential tower, the curvaceous forms creating a counterpoint to the otherwise linear road-facing southern façade.

The spacious entrance lobby is designed to draw people into the building, via a large portico structure. The apartment living areas face north with bedrooms and services areas orientated to the south.

The structural frame of the residential tower is almost entirely timber with steel connections, the floors are CLT panels, and the lift shaft constructed of light timber frame. The only non-timber elements to the structural system are the concrete slab ground floor and footings, and a number of small steel beams holding up the stairs. Mass timber construction is very much the hero of the project.

Context

Clearwater Quays is the leading project in residential mass timber in New Zealand. In 2010 and 2011, the Canterbury earthquakes killed 185 people and caused many billions of dollars of damage to the economy, with the collapse of unreinforced masonry structures all over the Canterbury region. Since then, understandably, there has been a significant uptake in the construction of seismically resistant buildings in the Canterbury region.

Clearwater is a resort centred around a lake next to a golf course, north of Christchurch in the South Island of New Zealand. A number of existing two- to three-storey 'boat houses' wrap around one part of the lake, while other features include a large clubhouse and restaurant. The project is distinctly upmarket and high-end luxurious, catering for those with a passion for golf. The Clearwater Quays apartment buildings will be the centrepiece of the development, not just for their height and architectural presence, but also for their environmental credentials, choice of low carbon materials and means of construction.

These apartment buildings have been carefully designed to have an excellent response to seismic action, while demonstrating the latest seismic-resisting principles available for use with mass timber construction. The development is a showcase demonstration project as part of the Ministry for Primary Industries 'Mid-Rise Wood' primary growth partnership programme.

Architect's statement

The entire structural frame was designed under the Structural Engineer's (Enovate) guidance. The methodology of the structure is that the LVL portals and beams provide bracing in one direction, whilst in the other direction, bracing is provided by the plywood lined light timber frame bracing walls.

Although a CLT slab was considered for the roof structure, timber trusses were chosen (to keep costs down), with rafters over the stairwell to gain additional presence and height. In addition to the concrete ground slab, the other exceptions to timber construction are a series of steel beams in the stair-well, used to reduce the depth of the original LVL beams. Because of their smaller cross-section, they were able to be recessed into the stair. Steel-splice joints are used to join the timber H-frames together and anchor them to the floor slab.

As it proved impossible to use timber cladding up the entire height of the tower due to fire/spread of flame restrictions, we settled for use of Abodo Vulcan timber (thermally modified) cladding within the deck/balcony areas with the CLT floors acting as the fire separation, and for the ground level garaging and separate garages. The remainder of the wall cladding is Flashclad's extruded aluminium Dualbord and Euro-line cladding in vertical/horizontal orientations. The fire engineer was able to contend with council that the balcony areas were internal, and because these were also sprinklered, they were able to be clad in timber.

The most sustainable aspect of the building is the carbon sink of the building's structure which has the added benefit of lightness, compared to a typical concrete and steel structure, therefore requiring a smaller foundation design. Due to the structure and timber infill framing's pre-fabricated nature (and also including sheathing and ply bracing), the design results in a much faster construction time and speedier 'closing in'. The living spaces are conditioned via reverse cycle heat-pumps, negating the need for a central heating system. Natural ventilation has been used for cooling in lieu of artificial, with trickle vents introduced for make-up air (with acoustic baffles to control exterior noise). Low-E glass was originally to be used just for the bay windows which have the greatest potential for over-heating, but following an upgrade from the glass supplier, all windows will now be double-glazed and low-E roof-mounted solar PV panels supply the majority of the common area (lobby and stairwell) electricity demand.

Looking back, we feel that the design process could have been more collaborative, which would have been fostered more had the scope allowances for the design engineers been agreed early on, and could have produced an even better and more streamlined result for the project as well as for the individual engineer's outcomes.

The main thing that we found with working with mass-timber construction, particularly with a multi-residential purpose, was the almost criminal constraint of having to cover it all up with suspended ceilings and floating floors to achieve the necessary acoustic sound transmission ratings. We had to fight with the acoustic engineer to keep the LVL columns and beams exposed within the apartments (and these may yet still have to be covered up if the flanking noise path decreases the performance enough following the acoustic site testing). The structural engineer was happy enough for them to be exposed, as they were over-designed from a fire point of view, so that the structure is able to withstand half-an-hour's charring without needing further fire-retardant coatings.

The exposed CLT stairs were challenging, due to the way that the steel beams integrate with the lighting and sprinkler system. Originally as the beams supporting the stairs were designed as LVL timber, they were quite deep and clashed with the services and impeded head height, so switching these to steel solved the problem, although somewhat antithetical to the timber showcase of the building, but the view of the steel is very minimal as it is recessed within the depth of the CLT stairs.

Engineer's statement

Seismic challenges in New Zealand are extreme compared to other parts of the world, and these challenges lend themselves to using mass timber construction. As a result, New Zealand's responses are quite different from the rest of the world too. Clearwater Quays Apartments has a very advanced system capable of dissipating any seismic energy. Although a relatively small apartment building by North American standards, the Clearwater project exhibits a structural framework system that can work on a larger scale (Newcombe, 2011). In the north-west to south-east direction a series of light timber frame, plywood-clad shear walls take the lateral load, connecting back to the core. In the north-east to south-west direction, there is a system of seismic H-frames made from LVL, which allow large openings to create open plan living spaces with great views onto the nearby lake. The third part to this seismic resisting construction is to construct floors as very resilient diaphragms, strictly controlled, using 135mm thick three-layer CLT panels. The CLT slab floors are fixed securely down into the LVL beams to work as one highly resilient structural system.

The large LVL H-frames are prefabricated and assembled in the large purpose-built TimberLab factory and are delivered to site by truck fully assembled. In fact, all the timber elements from all the different suppliers arrive and are assembled seamlessly on site. CLT floors, having been derived from the same building model, are also fixed without issues onto the LVL H-frames and then the prefabricated plywood shear wall panels are stitched in as well. Interestingly, the walls to the lift shaft and stairwell are not CLT, but instead simply timber frame clad externally to the south-west with plywood, while the lift shaft is lined internally with plasterboard. This is to ensure the seismic resisting system (the LVL frames) and the core respond in a compatible manner, preventing the entire structure from twisting severely during an earthquake. The stair landings themselves are manufactured from 160mm thick four-layer CLT.

Concision are the contractor for the fully prefabricated external walls as well, where panels are factory-fitted with plywood and Gib Weatherline sheathing panels, arriving on site fully assembled and ready to be quickly installed in place. Red Stag Timber framing was selected for its dimensional stability and carbon neutrality. The Red Stag mass timber factory (NZ) had not yet opened for CLT manufacture, but would have preferably been used to achieve a lower carbon footprint.

Inter-tenancy walls at Clearwater do not use the NZ-standard twin stud wall construction, but instead have a slightly different take on it. A 140mm thick timber frame wall is clad with plywood to form a rigid impenetrable layer, then a fully separate acoustic wall mounted on resilient rails is installed alongside, finished in layers of plasterboard.

The key difference with this project is the extensive use of a factory-prefabricated lateral load (wind and seismic) resisting LVL frame system. The frame system is an alternative to post-tensioned timber frame systems, also developed in New Zealand (see the research led by Professor Andy Buchanan as part of the Structural Timber Innovation Company) and avoids the use of specialist/high risk post-tensioning on site. Further, the high-performance frame system is expressed internally within the apartments, which demands that the structural connection detailing is discrete and attractive. To achieve the aesthetic and structural requirements, mild steel factory-installed dowels are applied in the critical beam-column joint regions. These dowels dissipate energy during a seismic event, similar to reinforcing steel in a reinforced concrete beam, but if there is a very large seismic event (with a return period of, say, 1 in 2,500 years) it is expected that the frame system may undergo some damage and require replacement (i.e. the system is not a damage avoidance seismic design). For this element, a middle ground has been reached between aesthetics and seismic resilience that is consistent with the plywood walls in the other direction (Figures 16.3–16.9).

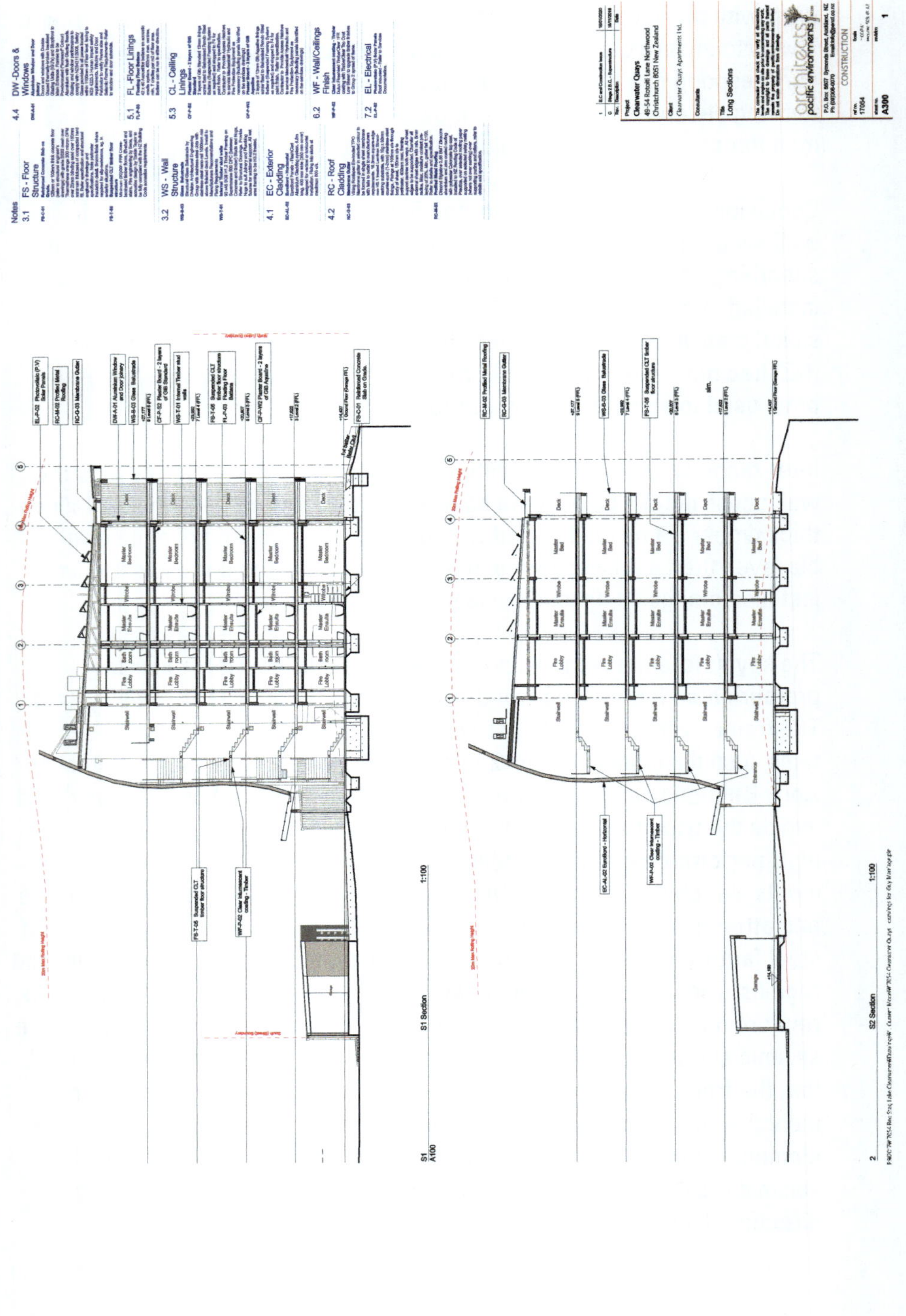

Figure 16.3 Clearwater Quays long section.
Source: Courtesy of Pacific Environments NZ.

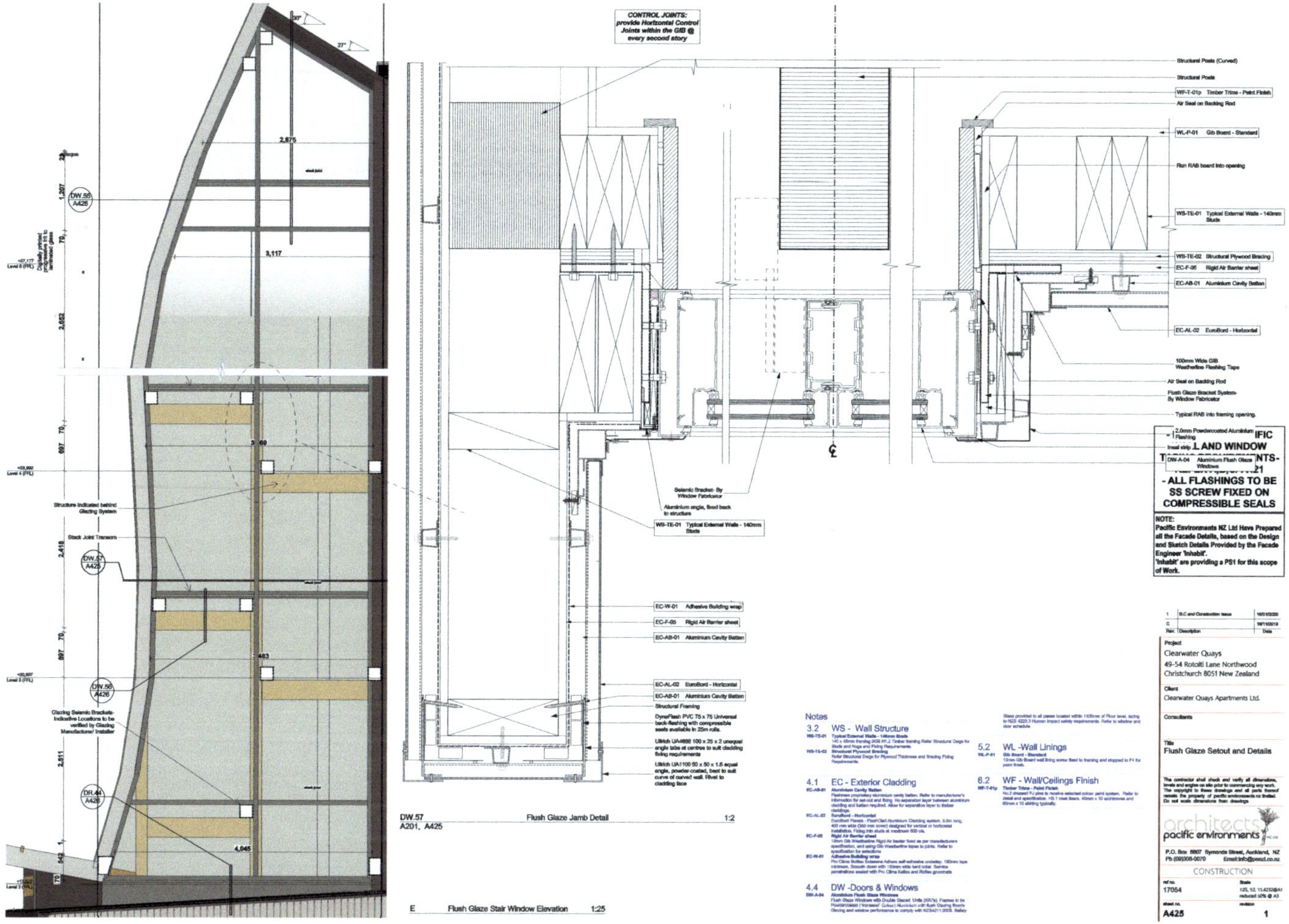

Figure 16.4 Clearwater Quays stair core.
Source: Courtesy of Pacific Environments NZ.

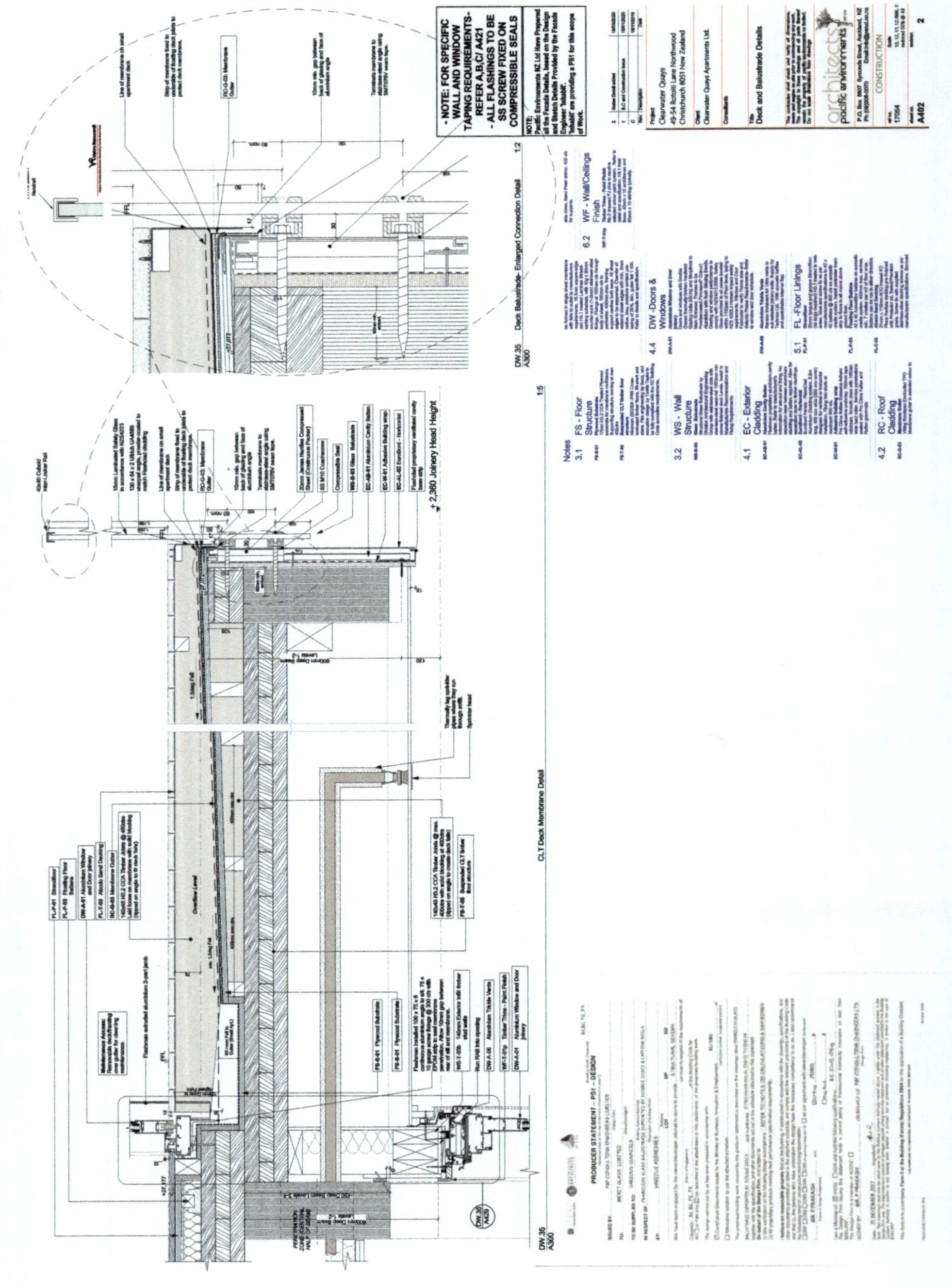

Figure 16.5 Clearwater Quays deck detail.
Source: Courtesy of Pacific Environments NZ.

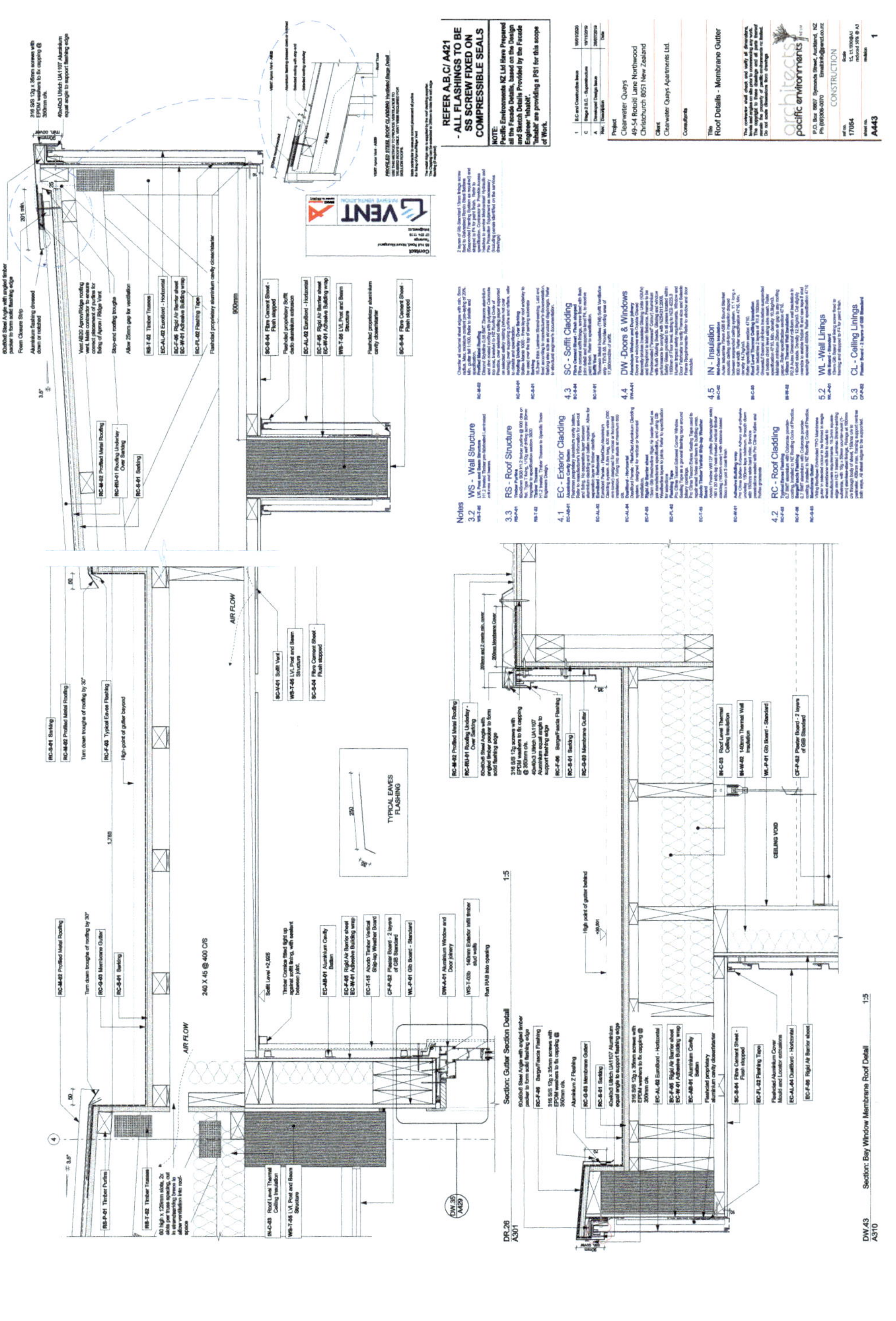

Figure 16.6 Clearwater Quays roof detail.
Source: Courtesy of Pacific Environments NZ.

Figure 16.7 Clearwater Quays 'The Gap'.
Source: Courtesy of Pacific Environments NZ.

Figure 16.8 Clearwater Quays ensuite.
Source: Courtesy of Pacific Environments NZ.

Figure 16.9 Clearwater Quays kitchen and living room.
Source: Courtesy of Pacific Environments NZ.

Works cited

Buchanan, A., Deam, B., Fragiacomo, M., Pampanin, S., and Palermo, A. 2008. "Multi-storey prestressed timber buildings in New Zealand." *Structural Engineering International, Journal of the International Association for Bridge and Structural Engineering (IABSE)*, 18, 2.

Cattanach, A. and Pampanin, S. 2008. "21st century precast: the detailing and manufacture of New Zealand's first multi-storey PRESSS building." *Proceedings of the 2008 New Zealand Concrete Industry Conference*, Rotorua, New Zealand, pp. 2–4.

Newcombe, Michael. 2011. "Seismic design of post-tensioned timber frame and wall buildings." PhD thesis. University of Canterbury, NZ.

Newcombe, M.P., Pampanin, S., Buchanan, A., and Palermo, A. 2008. "Section analysis and cyclic behavior of post-tensioned jointed ductile connections for multi-storey timber buildings." Journal of Earthquake Engineering, 12(Special Issue), 83–110.

Palermo, A., Pampanin, S., Buchanan, A., and Newcombe, M. 2005. "Seismic design of multi-storey buildings using laminated veneer lumber (LVL)." 2005 New Zealand Society of Earthquake Engineering Conference, Wairaki, New Zealand, p. 8.

17

Chapter Seventeen

Case study 2: Wynyard Central East 2

Case study 2: Wynyard Central East 2

Patrick Clifford

DOI: 10.4324/9781003123873-17

Figure 17.1 Wynyard Central schematic diagram.
Source: Courtesy of Architectus.

Project Title: Wynyard Central East 2 (Figure 17.1)

https://architectus.co.nz/work/wynyard-central/

Team info

Client: Willis Bond & Co

Architect: Architectus

Architect Team: Design Director: Patrick Clifford. Project Architect: Frank Coleman

Project Team: Alisha Patel, Amelia Moginie, Callum Barnett, Elizabeth Seuseu, Hamish McPhail, Jeremy Thompson, Kishan Patel, Mary Henry, Michael Whiteacre, Severin Soder, Warren Nicholson

Builder/Contractor: LT McGuinness

Artist: Maaka Potini of Ngati Tamaoho

Structural, Hydraulic, Mechanical, Electrical, and Façade Engineer: Mott MacDonald

Landscape/Planting: Architectus in Association with Panuku Development Auckland and its Iwi Liaison Group

Quantity Surveyor: Barnes Beagley Doherr

Surveyor: Yeomans Survey Solutions

Geotechnical Consultant: Tonkin + Taylor

Traffic Consultant: Traffic Planning Consultants

Planning Consultant: Barker & Associates

Fire Consultant: Aurecon

Acoustics Consultant: Marshall Day

Civil Engineer: Beca

Sustainability Consultant: eCubed

Engineer to the Contract: PwC

Project info

Summary

Wynyard Central is a predominantly residential mixed-use development on Auckland's waterfront which is undergoing a transformation from an industrial area into a vibrant inner-city neighbourhood. Architectus has been involved with the Wynyard Quarter for the past 20 years including authoring its Urban Design Framework; a fundamental premise of that document is to build on the character of the Wynyard Quarter into the future. The challenge with adding to the neighbourhood is to respect the character imbued into the precinct by the marine and fishing industry and the industrial quality that exists in many of the heritage buildings that are being retained. Wynyard Central adds to and complements this character.

Location:

141 Pakenham Street West, Auckland, New Zealand

The project comprises:

- 18,000m^2 floor area (excluding basement and laneways)
- 113 residential units
- 800m^2 food and beverage and retail space
- 2 laneways

Number of storeys:

- Apartments: 11 Storeys
- Townhouses: 3 Storeys
- Pavilions: 5 Storeys
- (Excluding 2 levels of basement car parking)

Site coverage:

- Site Area: 4,000m^2
- Site Coverage: 74%

Net area per floor and dwelling units per hectare:

- 282 dwellings per hectare
- 590 bedrooms per hectare

Number of apartments:

- 21 × 1 Bed
- 61 × 2 Bed
- 31 × 3 Bed

Average size of apartments excluding balconies:

- 1 Bed: 57m^2
- 2 Bed: 103m^2
- 3 Bed: 152m^2

Awards received:

- NZIA Sir Ian Athfield Award for Housing 2019
- NZIA New Zealand Architecture Award – Multi-Unit Housing 2019
- PCNZ Property Industry Excellence Award – Multi-Unit Residential 2019
- NZIA Auckland Architecture Award – Housing Multi-Unit 2019
- Home of the Year – Multi-Unit Category 2019 (Figure 17.2)

Context

Wynyard Central project.

In 2013, following an international design and tender process, Willis Bond was selected as developer of the residential lots in Wynyard Quarter. Architectus was part of this successful team together with Athfield Architects and Studio Pacific Architecture. The architects collaborated on the overall site strategy and subsequently took responsibility for individual components.

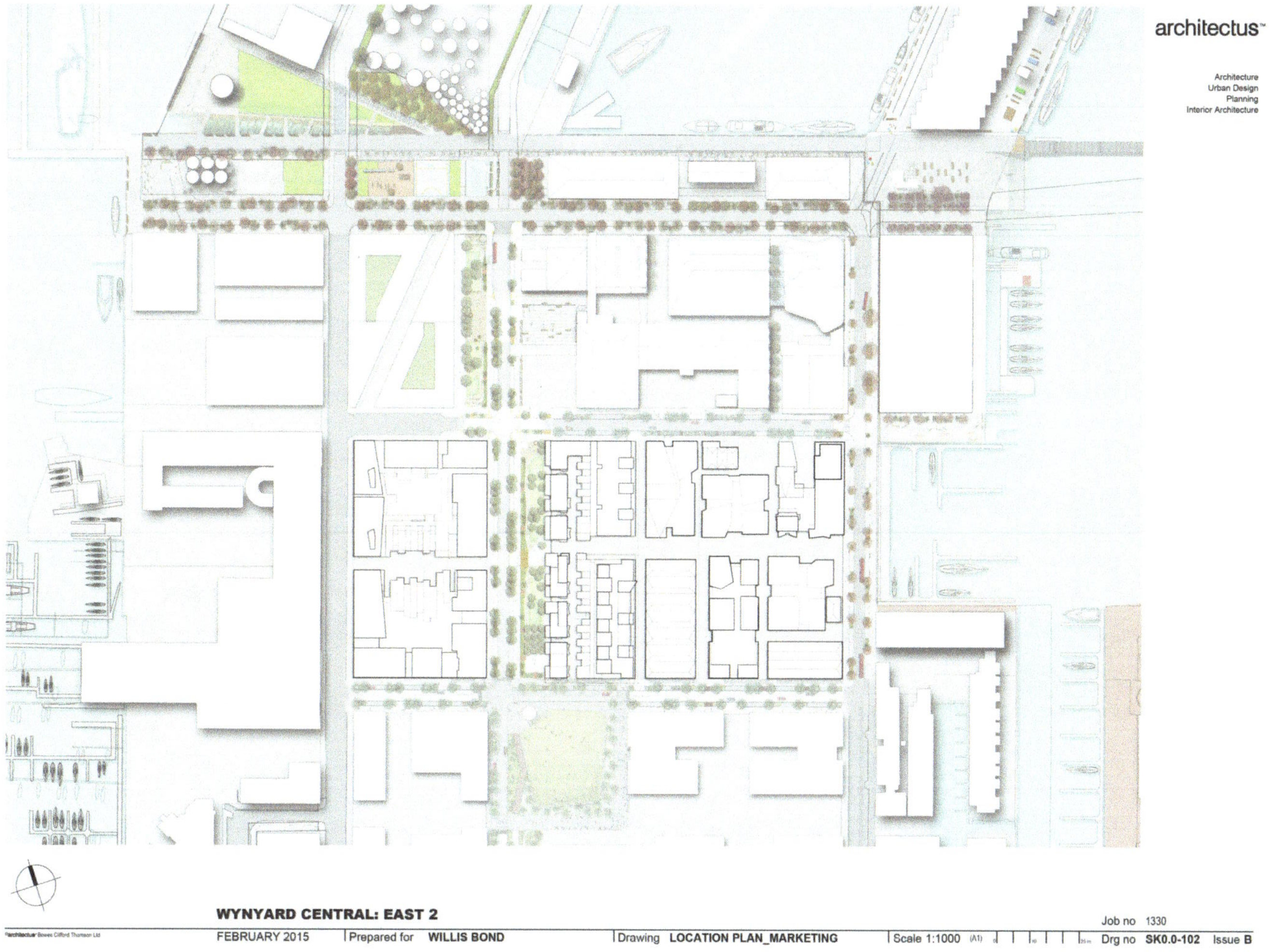

Figure 17.2 Wynyard Central location plan.
Source: Courtesy of Architectus.

The Wynyard Central project is part of an integrated, cohesive master plan. The plan's ambition is to bring the outside in, connect to spaces and uses within the wider waterfront precinct, and to form neighbourhoods within neighbourhoods. Housing has been designed for a variety of residents, with an emphasis on the development of family residences, through considered space planning as well as child-friendly amenities and safe outdoor spaces. Sustainability principles permeate every aspect of the design. Public spaces are celebrated, and ground floor edges are activated and engaging. Wynyard Central is home to a mosaic of lives and activity, interwoven and independent, in a thriving waterfront context.

Brief

The redevelopment of Wynyard Quarter on Auckland's waterfront is the largest transformational commercial and residential development opportunity in central Auckland. Prominently and strategically located to the west of the current CBD, Wynyard Quarter incorporates some of the best mixed-use waterfront land of any major city in New Zealand. Planning for redevelopment commenced in 2004 with several major infrastructure, workplace and hospitality projects established by 2010. A unique and exciting opportunity existed for the next stage of development in the Wynyard Quarter Central Precinct.

Eke Panuku (Auckland Council's Urban Regeneration Agency) had identified three strategic zones for redevelopment and sought private sector participation for:

- Commercial office development centred around an innovation hub
- International hotel
- Residential, including mixed-use apartment and townhouse developments.

Concept

'Walking in the shoes of a community' – creating a place for people

- The vision is for a diverse and lively community
- A variety of residential typologies
- A diversity of residents
- Access to a range of outdoor spaces
- Different neighbourhoods with their own distinct atmospheres
- Variety of scale and materiality
- A mixed-use community (curated food and beverage and retail tenants)

Past and future inter-twine to provide a rich experience of urban village life for residents of different ethnicities, age groups and occupations.

These diverse residents share a desire to live in this unique place between the inner city and the waterfront. The master plan for the five residential precincts imagined a diverse and lively community in which residences, retail workspaces, public and private spaces, and amenities are woven together. This collaboration created residential spaces unique to the Quarter, homes that reflect the requirements of Auckland, and the diversity of its population. The variety of residential typologies, with their supporting cast of public and private spaces has attracted different household configurations: singles, couples and families at various life stages.

The diversity of the residents is reflected in the varying scale and materiality of the buildings. In scale, one, two and three-bedroom homes have been created in buildings ranging from 3 to 11 storeys, with diversity in materiality including brick, wood, glass and concrete. All residents have ready access to the laneways, public spaces, waterfront community amenity and private spaces in the neighbourhood as well as the lifestyle opportunities provided by Auckland's inner city and waterfront.

A large part of the rationale of this diversity is to make the area accessible to all Aucklanders and not just for a small elite; to acknowledge and respond to the housing shortage in Auckland and bring a wider cross-section of the city's population to live on its waterfront. The master plan is critical in delivering diversity – it supports the delivery of a curated collection of bespoke buildings of different scales, materiality and different typologies. Each building is orientated to optimise sunlight, views and a sense of community and place.

Master plan

The Master Plan for the W2 +W2, E1 +E2 and S7 is one of integration, but also of difference. In short, it incorporates:

- Design informed by best practice but bespoke to the Wynyard Central community and neighbourhood.
- Sustainability embedded into the precinct from day 1.
- A respect for place and people through design which acknowledges the diverse population of the city, fosters community, and references the precinct's connection to Auckland's marine environment.
- Abundant open public space and amenities including the Daldy Street Linear Park, Amey Daldy Park and ultimately Headland Park. This integrated framework of public spaces between the built forms is designed to bring 'life to the streets' of the village.
- Houses with their own public and private open spaces.

- Building heights that vary from 3 to 11 storeys, each building designed to complement neighbouring buildings both within the precinct, and the broader neighbourhood.
- Pedestrian and cyclist-friendly streetscapes and laneways that connect between neighbouring streets, mediating traffic flows to the advantage of residents and visitors.

Further guiding master plan principles:

The great outdoors

Focus has been given to providing useable outdoor living spaces as an extension of the apartments – something very important in Auckland's climate, with cross-ventilation wherever possible. Buildings always have appropriate edges to the streets.

Lanes and public spaces

The lanes are key to the place-making of this new community. They provide accessibility for pedestrians and vistas of the surrounding neighbourhood. It is the integration of the buildings and their interconnecting lanes that provides the surprise, intimacy and energy of the public spaces. In these areas, the public and private meld on the streets, where the retail and public amenities meet the entrances to the private homes in a shared space.

Servicing and parking – traffic flows

The design acknowledges the proposed hierarchy of circulation for pedestrians, vehicles and bicycles around the perimeter and through the public spaces. The urban structure of Wynyard Central's residential areas is comprised of eight city blocks separated by a grid of streets for pedestrians, cyclists and vehicles. The proposed street network includes the extension of Daldy Street in a north-south direction through the block bound by Madden Street and Pakenham Street and the insertion of a series of public lanes to heighten the accessibility and walkability of these blocks. Halsey and Beaumont Streets are primarily for vehicles. Pakenham and Madden Streets allow access for service and resident vehicles into the Wynyard Central parcels. The east-west pedestrian street is a key amenity circulation route, and provides framed views through the blocks and parcels to the Viaduct Basin in the east.

Transport planning

Wynyard Quarter's transport strategy is based on a 70%/30% modal split in favour of public transport. Consequently the intention at Wynyard Central was to encourage fewer car trips within the quarter and more walking and cycling. The design makes cars a less convenient option and reduces the impact of their use on people visiting, living and working in the precinct.

Car parks are disconnected from apartment ownership in order to provide flexibility for residents, and the provision of bike storage has been prioritised. Walking and cycling pathways connect naturally with the much larger promenade that will stretch from the Wynyard Quarter to the Harbour Bridge.

Sustainability

The redevelopment of Wynyard Quarter represents a unique opportunity in central Auckland – one which encourages integrated planning and design processes, and one which encourages planners, designers and builders to meet the challenges of a post-carbon, climate responsible world. Passive design was a key focus, with the overarching design intent to make the residences largely independent of a need for artificial heating and cooling for 80% of the year and independent of a need for artificial lighting in their habitable spaces for 80% of annual daylight hours.

Other sustainability attributes include:

- heat-recovery ventilation systems and heat-pump heating
- high standards of thermal performance, daylight and sound insulation
- high level of acoustic privacy
- 95% LED lighting
- smart metering for apartment energy and water
- rainwater harvesting for selected homes and common uses
- WELS-rated water fixtures and ENERGY STAR-qualified appliances
- custom double-insulated hot-water cylinders
- materials are low VOC, and certified by ECNZ or another approved eco-label
- shared hybrid-powered cars available for residents' use
- bicycle parking for residents and visitors
- low-speed, shared streets and laneways for walkability and cycling
- low-impact design, rain gardens, stormwater treatment
- resilient design, renewal and rejuvenation of previously underutilised land
- 70%–90% of demolition and construction waste diverted from landfill
- recycling/composting facilities
- green roofs and linkages to public green space including linear park and Wynyard commons
- connection to WQ Smart precinct providing feedback on the environmental performance of the precinct as a whole.

Resilient design and adapting for climate change

Resilient design plans and designs for redundancy and durability of critical infrastructure and systems during extreme events are needed – such as those expected to be more frequent due to climate change: sea level rise, storm surge and more extreme rainfall and winds. Climate change and sea level mitigation micro-strategies adopted for the Wynyard Central Residential Developments include:

- Adopting energy efficient and low greenhouse gas emission systems within the apartments
- Having no fossil fuel (e.g. gas) supply to the apartments
- Setting of an appropriate residential ground floor level with respect to flooding and sea level rise. The residential floor level of Wynyard Central East 2 sits at a minimum of 1m above ground level
- Sizing stormwater systems with a 15.1% increase in rainfall due to climate change as per Auckland Council Code of Practice for Land Development and Subdivision
- Sizing heat pumps for 1°C warmer in summer than the NIWA design recommendations
- Reducing the impact of basement car parking to one level by the use of above ground sleeved parking
- Protecting the entries to the basement car parking with a facility to add storm boards
- Locating critical infrastructure such as electricity substation at ground level on plinths rather than in the basement
- Providing both active and passive environmental features to the apartments so that they can still function adequately in the event of a power outage following an extreme weather event
- Adopting wind tunnel testing to test, and where necessary, mitigate the wind effects on the pedestrian environment. Also, to inform the façade design parameters.

These strategies mitigate and protect residents and visitors to the precinct from the anticipated adverse effects of Climate Change and Sea Level Rise (CCSLR) and ensure that the built infrastructure is robust and resilient to anticipated meteorological and climatic events. Strategies to reduce and control stormwater, and to provide resilience against extreme sea levels, will ensure that a safe and habitable environment is provided for residents, even during extreme conditions. They will also minimise the risk of pollutants being released into the natural environment during extreme events.

Homestar

Homestar is an independent certification tool that evaluates the standard of environmental performance of homes in terms of energy efficiency, health and comfort, water, waste home management and more. A home is rated on a scale from 6 to 10.

Wynyard Central is one of the first two residential developments in Wynyard Central Precinct. As part of the Wynyard Central Sustainability Standards, it is a requirement that all residential developments achieve a minimum seven Homestar rating. These first two projects have been instrumental in the development of Homestar to be jointly applicable to multi-unit apartment type residences through both sponsoring and piloting Version 3 of the Homestar tool. Much of the feedback from these projects has also been subsequently incorporated into the most recent update of Homestar, Version 4.

The final Homestar Built Ratings have recently been awarded for the apartments at Wynyard Central. All the apartments are rated at either seven or eight stars. Being a pilot project for the new version of Homestar has made this a challenging process and the result is a fantastic achievement and reflective of the NZ leading sustainability aspirations of the project from concept all the way through to delivery on site.

Wynyard Central

– 113 dwellings, 29 design types – 77%/23% 8/7 star design split

Pavilions

– 25 dwellings, 8 design types – all 8 star design

Construction info

Structure

The structure is founded on cast in-situ concrete piles extending down through historic reclamation fill into the bedrock below. These piles support a two-way cast in-situ reinforced concrete slab/foundation beam system to form the basement floor. Precast concrete basement walls contain the perimeter of the site. The basement and ground floor structures form the podium from which the apartment tower and the three separate Pavilion structures extend.

From the ground floor/podium level, two-way structural steel beam-and-column frames extend the full height of the building. The frames support precast concrete flooring units with in-situ concrete topping. Lateral loads are taken to in-situ/precast concrete vertical circulation shear cores. Three cores are provided for the 11-storey apartment block, and one core for each of the three Pavilion buildings. The roof structure consists of structural steel frames with lightweight steel purlins and steel rod cross-bracing. 150mm secondary timber wall framing provides support to both the Pavilion vertical cedar cladding, and the Townhouse brick veneer cladding.

Enclosure roofs

Warm roof systems are incorporated throughout the project. The system comprises a single layer KEE membrane installed over a high-density acoustic insulation board, and 80mm thick PIR foam thermal Insulation board over a proprietary long run metal roofing undertray. The warm roof system build up spans across lightweight proprietary galvanised steel purlins.

In limited areas where the concrete slab forms the roof structure, the metal roofing undertray is not required and the warm roof system is adhered directly to the slab.

Green roofs

The Pavilions and Townhouses incorporate green roofs over the warm roof system. The green roofs provide enhanced thermal and acoustic insulation, assist in moderating interior environment fluctuations, and filter rainwater which is captured for secondary use.

Proprietary pre-grown planting modules are installed over filter cloth and 30mm rigid cellular drainage/roof membrane protection sheet.

The planting modules incorporate selected plant species native to the area and suitable to withstand the particular roof top microclimate. The growing medium is tailored for the local climate to assist plant growth and to retain moisture. Roof areas not covered by green roof planting are covered with river stone pebbles to provide a landscaped finish to the membrane and offer additional UV protection.

Decks

The Pavilion and Apartments decks are typically a concrete structure laid to falls, and Townhouse decks are treated plywood on timber joist structures. The deck membrane comprises a single layer KEE membrane installed directly on to the concrete or ply substrate. Pavilion decks are finished with 90 × 20 FSC certified hardwood fixed to timber joists supported on adjustable polypropylene deck jacks. Apartment and Townhouse decking consists of 600 × 600 × 20 engineered porcelain tiles also supported on deck jacks.

Windows

Window systems are combination of 40mm commercial powder-coated aluminium windows together with 160mm commercial flush glaze powder-coated aluminium sliding doors suites. Glazing generally consists of Insulating Glazed Units (IGU) throughout the project. The IGUs incorporate high performance glass for solar control, thermal insulation and acoustic insulation.

External cladding

The pavilions are clad with western red cedar vertical boarding on cavity battens. This system is installed over a self-adhesive vapour permeable water-resistive building wrap system, on a rigid air barrier (RAB) board on to 150mm timber framing. The cedar is band-sawn finish and coated with a penetrating stain.

The Townhouses are clad with a 70mm clay brick veneer over a cavity on to a self-adhesive vapour permeable water resistive building wrap system, over a RAB board on 150mm timber framing.

For the Apartment building, ground floor wall cladding consists of 70mm clay brick over a cavity, PIR insulation and precast concrete panels. Above ground, cladding is generally 150mm precast concrete wall panels with a high build acrylic paint finish. A timber inner wall framing with polyester insulation completes the wall build up.

At the Penthouse level 2mm PVDF coated aluminium panels are installed over cavity battens on to a self-adhesive vapour permeable water resistive building wrap system, a RAB board and 150mm timber framing.

Throughout the project flashings, cappings and trim are generally 1.6/2.0mm powder-coated aluminium with prefabricated corners and junctions for accuracy, robustness and weather tightness.

Planters and balustrades

Planters are installed to all decks and integrated into the overall balustrade design strategy. The planters are constructed from 8mm welded aluminium plate with a powder-coated finish and are bolted to the structure through stainless-steel uprights.

Powder-coated aluminium was selected to provide accurate fabrication, long term durability, low maintenance and light weight. All planters were prefabricated and bolted into position as fully completed units. Prefabrication provided the benefit of accurate and consistent quality together with offsite construction programme advantages.

As for planters, the Pavilion balustrades are prefabricated from welded aluminium flat bar and powder-coated. Apartment balustrades consist of stainless-steel uprights and continuous stainless-steel handrails. Stainless steel was selected to provide long term durability with low maintenance.

Stainless-steel spider fittings from the uprights support toughened laminated glass infill panels. As with the planters, the stainless-steel balustrade system was prefabricated to gain quality and construction programme benefits.

Screens

Prefabricated timber screens are installed in the Pavilion buildings.

The fixed vertical screens consist of glue-laminated timber fins complete with predrilled holes and slots for bolt fix installation into prefabricated structural steel supports.

The sliding screens incorporate prefabricated glue-laminated timber fins bolted in to powder- coated aluminium sliding door frames with aluminium running gear.

Angled window eyebrow screens are constructed from 30mm thick western red cedar slats screw -fixed to prefabricated powder-coated stainless-steel frames.

Interior internal partitioning

Internal partitioning throughout the project is generally steel stud with plasterboard lining.

Party walls between apartments are double-framed, insulated and lined full height to achieve the required inter-tenancy fire and acoustic ratings.

Ceilings

Ceilings are a suspended steel proprietary system with flush stopped plasterboard linings. The ceiling cavity is insulated with polyester insulation, and combined with the concrete slab above, is designed to meet inter-storey acoustic requirements.

Floors

With the exception of the Townhouses, floors are concrete. Floor coverings are a combination of carpet on underlay, engineered timber flooring planks laid over an acoustic underlay, and ceramic tiles over acoustic underlay and waterproofing. The concrete slab provides the inter-storey fire rating. The inter-storey acoustic rating is achieved with the combination of floor coverings and underlay, the concrete slab and the insulated plasterboard ceiling below (Figures 17.3–17.10).

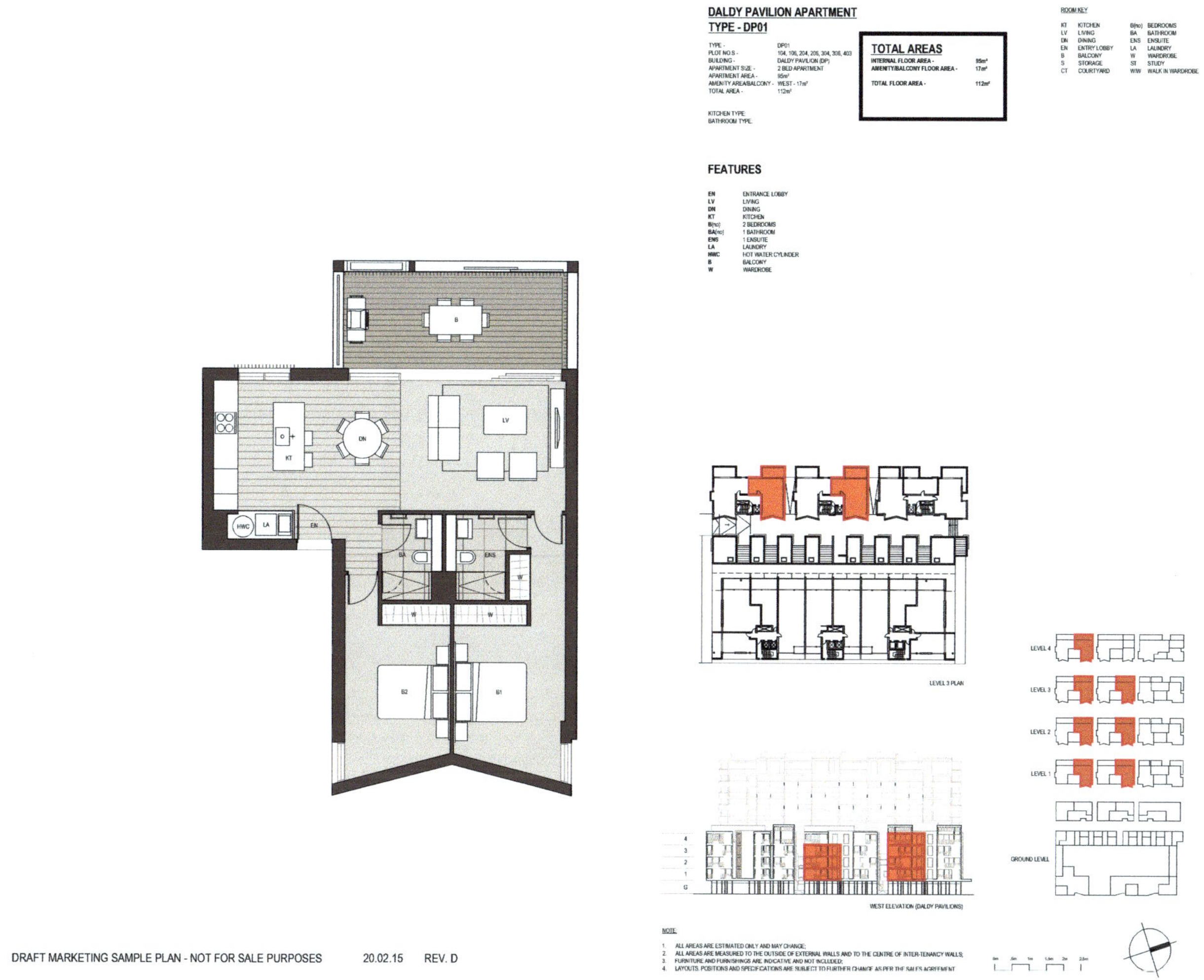

Figure 17.3 Wynyard Central type 1 apartment plans.
Source: Courtesy of Architectus.

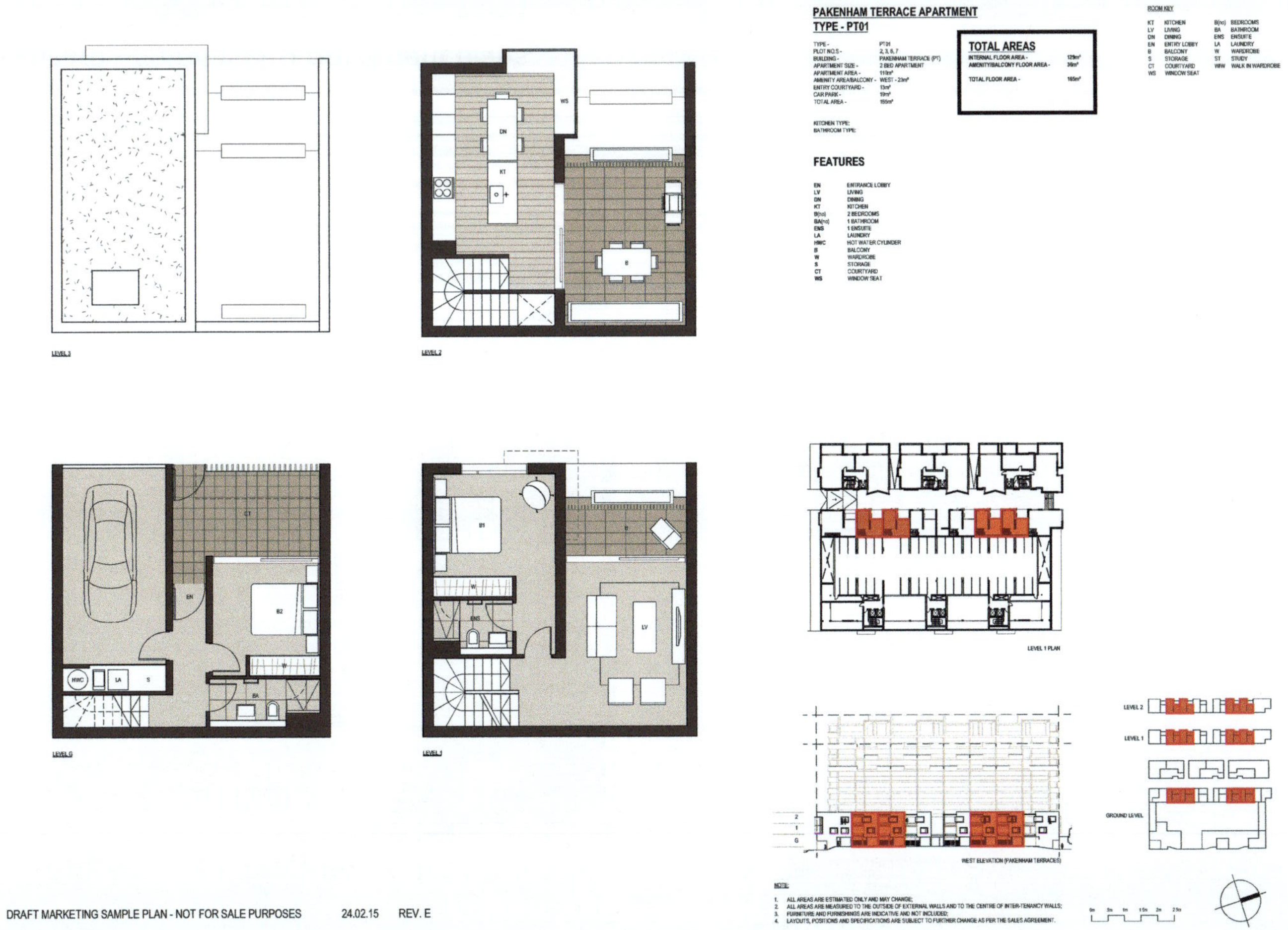

Figure 17.4 Wynyard Central Terraces apartment plans.

Source: Courtesy of Architectus.

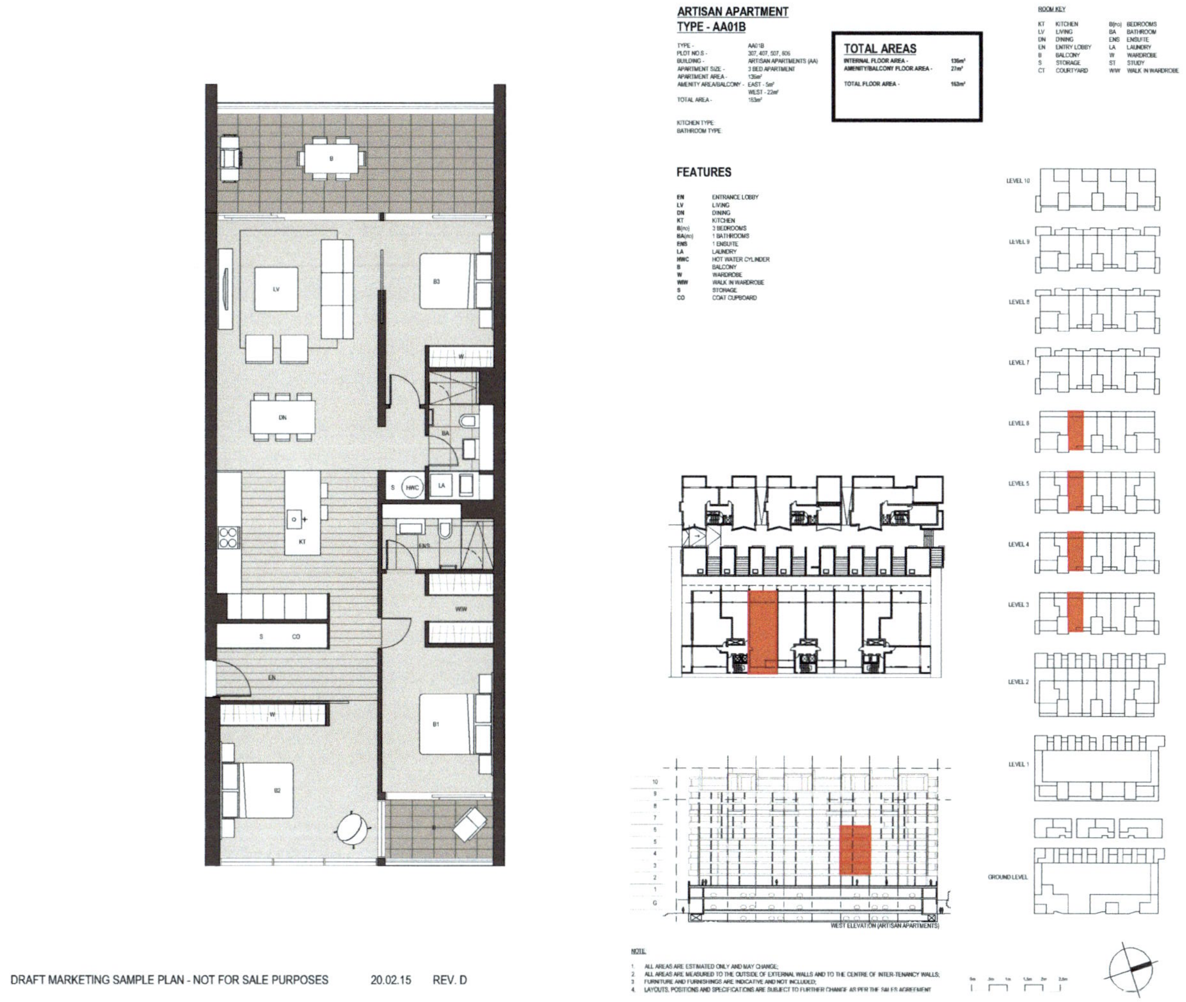

Figure 17.5 Wynyard Central Artisan apartment plans.
Source: Courtesy of Architectus.

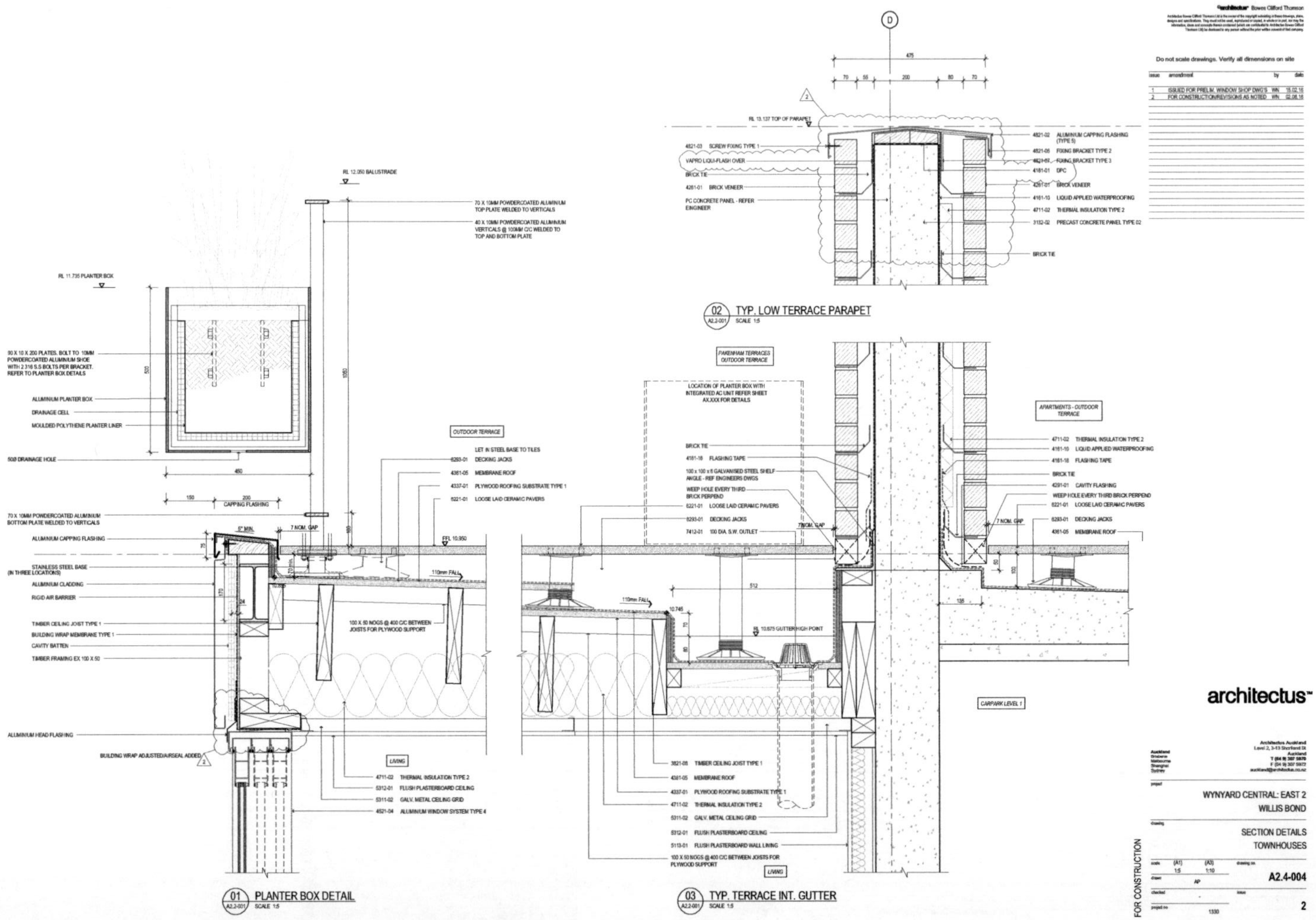

Figure 17.6 Wynyard Central deck and planter detail.
Source: Courtesy of Architectus.

Figure 17.7 Wynyard Central view into Mews.
Source: Courtesy of Architectus. Photo by Simon Devitt.

Figure 17.8 Wynyard Central vista along street front.
Source: Courtesy of Architectus. Photo by Simon Devitt.

Figure 17.9 Wynyard Central East 2 end elevation.
Source: Courtesy of Architectus. Photo by Simon Devitt.

Figure 17.10 Wynyard Central – view back to the city.
Source: Courtesy of Architectus.

18

Chapter Eighteen

Case study 3: Dalston Works

Case study 3: Dalston Works

Kieran Walker and Guy Marriage

DOI: 10.4324/9781003123873-18

Figure 18.1 Dalston Works complete, with the London Overground rushing by.
Source: Courtesy of Waugh Thistleton Architects. Photo by Daniel Shearing.

Project title: Dalston Works (Figure 18.1)

http://waughthistleton.com/dalston-works/

Team info

Client/Developer: Regal Homes

Architect: Waugh Thistleton Architects (WTA)

Architect team: Andrew Waugh, Dave Lomax, Kieran Walker

Engineer (CLT): Ramboll (Denmark)

Engineer (Concrete): Pringuer-James Consulting Engineers (PJCE)

Engineer (Mech & Electrical): XCO_2 Energy

CLT contractor: B&K Structures (Derby, UK)

CLT supplier: Binderholz (Austria)

Date of construction: 2017

Project info

Summary

The concept for Dalston Works was to take a brownfield urban site in London and create a vibrant new community in buildings largely created from cross-laminated timber (CLT). With a total over 121 apartments in various sizes, the Works has been broken into several different volumes to reduce the bulk and create pleasant urban living spaces. The project in Dalston has been carefully designed to maximise daylight into the central courtyards, affording good daylight to all residents. It is the world's largest CLT apartment building, and is a landmark project rolling out the use of timber construction in high-density urban housing, across London and beyond.

The housing development is flanked by 1,500m^2 of retail and restaurant space, while to the south of the site, there is a flexible workspace hub for the burgeoning creative community within Dalston. There are two medium-rise blocks (a four storey block of offices, and a six storey block of apartments), that comprise phase one of Dalston Works, while phase two has a ten storey apartment block, a total of 33m high. It was the tallest CLT building in the world when completed in 2017 (Schuler, 2018).

Location:

67–71 Dalston Lane, Hackney, London.

Number of storeys:

Five to 10 storeys: one floor basement, then:

- Ground floor retail plus four floors of offices
- Ground floor retail plus six to ten floors of residential.

Gross floor area:

14,400 m^2

Number of apartments:

121 total

Approx. construction cost:

Not available

Awards received:

- AJ Sustainability 2017 (Highly commended)
- Structural Timber 2017 Solid Wood/Housing/Overall Winner
- NLA 2017 Housing (Commended)
- NLA Ashden 2017 (Highly commended)
- Offsite 2017 (Highly commended)
- RICS Awards 2018 (Shortlisted)
- Construction News 2018 (Shortlisted)
- Evening Standard Eco Living Award 2018 Winner
- BCIA Carbon Reduction Initiative of the Year 2018 (Shortlisted) (Figure 18.2)

Context

Sited near Dalston Junction railway station, Dalston Works builds on the urban bus and train links that exist in the East End and sits near the long-running Dalston street markets. Surrounding buildings are from the Victorian and Edwardian eras over 100 years ago, which are predominantly low-rise and constructed from London stock bricks, mixing with other newer apartment tower blocks and former industrial warehouses. The East London Line extension was routed through Dalston Junction after a long absence and the High Speed 1 (HS1) route and the future Crossrail 2 (CR2) line are also planned to connect through this point one day. With the Ridley Road street market nearby as well, Dalston Works is a classic example of transport-oriented development (TOD), rejuvenating a formerly rundown brownfield site.

Project design

Waugh Thistleton were the architects for the ground-breaking Murray Grove apartment building in Hackney back in 2009 (9 storeys, 47 apartments). Never before had CLT been used to such a large extent – it was the most adventurous use of CLT at the time, and proved to be an inspiration to many other architects around the world. The Dalston Works project was a chance for WTA to take their learnings from Murray Grove and apply them to a project that was almost three times larger. The London Borough of Hackney is home to both projects and is keenly supportive of fresh new ways of thinking to solve their housing shortage.

Kieran Walker, project architect on Dalston Works, notes that while CLT was a major feature, the project was so much more than that. Although the building is large, the volumes on the site were broken down to lessen the impact and the apartments centre around a couple of large courtyard

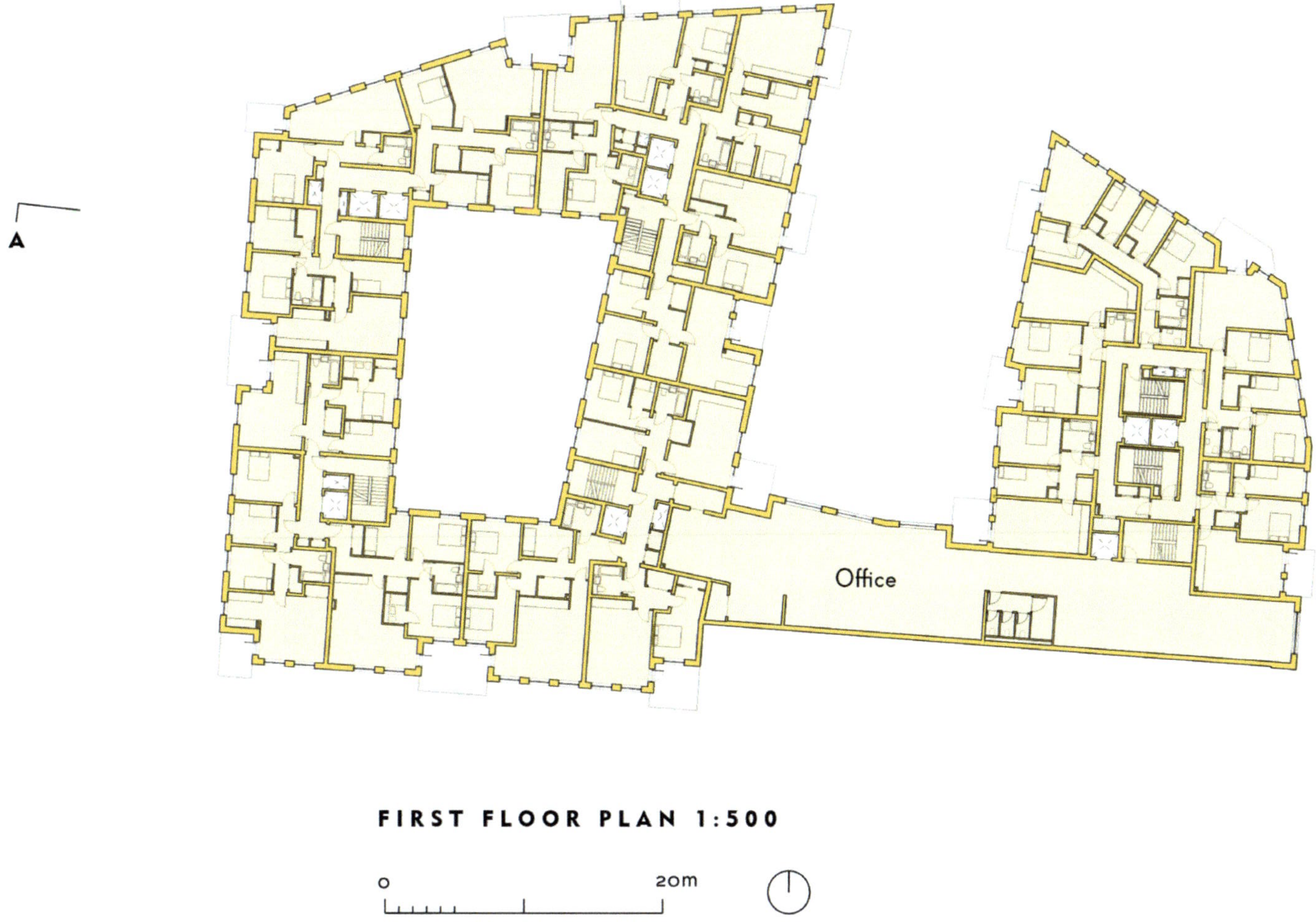

Figure 18.2 Dalston Works first floor plan.
Source: Courtesy of Waugh Thistleton Architects.

volumes opening off Martel Place (Walker, 2018). There are four parts to Dalston Works: low-rise office floors to the south, a block of shared ownership/affordable housing in one wing, a taller block of rental apartments in the other wing and finally the tallest block of apartments rising prominently in the centre.

It is estimated that a building of this size constructed from concrete would have weighed approx. 10,700 tonnes, far too heavy for the railway lines planned to run below. The distinct advantage of using CLT for this project was the drastic reduction in weight. The CLT structure weighs only 2,300 tonnes, thus allowing the client to also provide a further 35% more residential accommodation (Merrick, 2017) over several more floors. Rather than piling, which would have clashed with the future train routes for CR2 and HS1, a raft foundation of concrete was provided across the site on a thick bed of gravel. A thick (500mm) first floor concrete transfer deck takes the gravity load of the CLT and redistributes the loads down to the raft foundation below. The concrete deck also allows the timber superstructure to be started from a clean and dry substrate, set very level, away from the muck and mud of a usual building site.

The composition and massing is deliberately fashioned to tie in with the brick Victorian and Georgian housing and warehouses surrounding the site: the new brick-clad building settles comfortably into the solid brick surroundings. Nothing visible externally indicates that the structure of the buildings all have CLT at their heart. As parts of the building are well over the 18m height limit for timber buildings in London, the fire regulations have dictated that the structure needs to attain 90/90/90 FRR, everywhere except the stair cores – which are rated at a full 120/120/120. The timber structure gains the required ratings through a mixture of allowing for a certain amount of charring factor, but mostly by additional layers of plasterboard on walls and ceilings.

Sustainability

All the CLT panels were made and cut to measure in the Binderholz factory in Austria, and delivered daily to the UK on a 'nose to tail' (just-in-time) basis. Austrian forests are managed on a sustainable basis: an estimated total of 2,325 trees were used, producing a total volume of 4,649 m^3 and a calculated 3,576 tonnes CO_2e of sequestered carbon. All openings for windows, etc. were cut out in Austria in a process where even the sawdust was collected and turned into wood fuel pellets. The advantages of the CLT construction is not just the dramatically lower weight of the building but also the huge reduction in site deliveries – only 111 truckloads direct from the factory in Austria compared to an estimated 700 truckloads to complete the same sized building in concrete. No double-handling either, with panels coming direct from the factory to site, completely pre-cut to size, and ready to slot straight into site. That was of vital importance to a busy city location such as Dalston. Waugh Thistleton has calculated that by using CLT they have reduced CO_2 emissions equivalent to taking 1,703 cars off the road for a year.

Construction information

Spruce CLT BBS panels are used throughout this project, not only for floors, roofs and walls, but also for stairs and lift shafts. A ground floor podium was constructed in reinforced concrete to allow large open spans for ground floor retail, but above that all, the construction is CLT. The CLT panels are not visible however, either internally or externally, as they are faced externally with a vapour barrier, insulation and brick cladding – and lined internally with layers of plasterboard. A 100mm layer of foil-faced rigid insulation was fixed to the outside of the CLT wall panel, with a 90mm stainless steel angle bracket bolted to the CLT, taking the load of the brick masonry. To tie in with the local surroundings as much as possible, the entire development is then clad in a range of warm, red, full-size brick. The office building is similarly clad in bricks of a more grey colour.

All the window openings have been detailed in the manner of traditional brick detailing with projecting brick window sills: the visual appearance from the outside is purely that of a series of inter-related brick buildings. Window sizes are larger than a Victorian dwelling (more Georgian perhaps), well-proportioned with a mixture of fixed panes and opening sashes. The windows are a slight mixture of widths as well, avoiding monotony, as befits the needs of an apartment building. At the top of the parapet, the bricks are laid vertically in a soldier course, giving a definitive top to the wall.

The CLT panels were easily assembled on site by craning into position straight from the delivery trucks and fixed using self-tapping screws, angle-brackets and nail plates. Propping was not required, as it would be for a concrete slab building, while the quickly-installed CLT panels also provided a dry and weatherproof environment to work in. Panels used were typically 2.2m wide (comfortably transportable) and up to 13.5m long. Panels of 140mm thick CLT were used for the apartment floors, with wall panels varying in thickness: a full 160mm at lower levels reducing to just 100mm thick on the top floors. The building process was not only fast, but also quiet, just the rattle of battery-powered impact drivers fixing the panels together – no need for heavy tools. The number of workers on site can be fewer as well, with only four or five workers needed on site to assemble the CLT structure. The advantage of mass timber panels is apparent too in the fixing of both external cladding brackets and internal services – in both cases simply screw the fixing to the timber, and you are able to proceed with work sooner due to the quickly-enclosed working space.

CLT sizing was kept deliberately flexible at tender time so that any of the big three European suppliers (KLH, Stor-Enzo and Binderholz) could put forward a tender bid for the project. Binderholz Bausysteme (BBS) won the bid and their CLT prefabrication was utilised to the full by the architect and engineer team. The building's structural CLT panels were precisely machined by Binderholz in Austria, with window openings, vapour and fire barriers and M&E channels all planned out and installed prior to their arrival on site (Merrick, 2017). This approach speeds up erection on site and benefits the programme considerably, to the extent that it may ultimately reduce the management role of the main contractors. It also reduces the extent of wet trades on site and helps reduce the clash of multiple sub-contractors all working on the same project at the same time. Of course, the flip side of this is that the design team needs to have planned everything out on screen first: there's no room for change on site later.

B&K notes that the use of CLT on this project saved between 10% and 15% on the cost of the overall project, plus considerable savings in time. The six-storey block took just 12 weeks to assemble, while concrete would have taken twice as long. CLT panel cutting permitted other innovative time and cost savings as well, such as:

> CLT panels incorporate full-height window openings complete with temporary barrier-rails across the openings to help improve site safety. The barriers are formed by the manufacturer cutting a series of three holes, one above the other, to form the window, rather than a single opening. The holes are separated by two handrail-wide strips of timber which remain in the opening forming the barriers. When the time comes to fit the glazing, the temporary barriers will simply be cut from the window opening using a handsaw so that the glazing can be installed.
>
> (Pearson, 2016)

Floor spans for the apartments were kept to a maximum of 5m to allow the floor slabs to remain thin, while the assembly method created a rectangular honeycomb of individual interconnected rooms. With minimal seismic issues to be considered in London, the key lateral load condition was the force of wind:

> Ramboll... developed a connection detail to increase connection stiffness between the project's vertical wall panels, which will help limit the

> amount the building will deflect in the wind. The connection will also help prevent the floor panels being crushed by the weight of the structure above. The connector comprises four angle brackets. These are positioned above and below the floor panels on either side of the wall. The brackets above are linked to the brackets below by bolts. When the wind blows it causes one side of the connection to rise and the other to compress. The forces on the side that rise are transferred from the bracket through the bolt to the bracket below and then down through the wall. This arrangement avoids the structure pulling up on the floor panel. Between the floor panels a non-shrink, non-compressible grout plug is fitted to prevent the floor being crushed on the side of the connection under compression. The panels were subject to crushing tests at the University of Cambridge to confirm the connection's performance.
>
> (Pearson, 2016)

These grout pockets also help counteract any tendency for the timber in the building to undergo shrinkage, with a predicted 60mm shrinkage overall in height (without grout pockets) reduced down to just 20mm over the full height by using the grout. This small amount of shrinkage is easily absorbed within the 10mm movement joint at each floor of brick cladding (Walker, 2018).

There are several circulation cores in Dalston Works – one core in the lower building, another serving the office block, and four cores surrounding the open courtyard of the tallest building. Each core houses one, sometimes two lifts, with an escape stair tucked in nearby. Internal corridors are short and simple, while apartments are a mix of sizes and aspects: some single-aspect, some dual-aspect, and some corner-aspect. Floors are not completely rectangular, with slight shifts in the façade reflecting the angles of the surrounding streets. There is considerable variety in the apartment planning and as we all know, variety is the spice of life.

One wall of the development was sited close to an existing building, so there was no room to apply cladding to the wall at a later stage. To allow for this, this particular wall was composed of prefabricated CLT panels complete with external insulation and weatherproof cladding, lowered into place and fixed fully finished to the outside: far easier than if the construction had been steel or concrete.

The building features external balconies to nearly every apartment: some recessed, some expressed. The balconies are a design choice that required careful detailing from the architects and engineers. Created from steel with all balustrading relentlessly slim, black and vertical, all the balconies were installed fully assembled and fixed via multiple brackets (around five brackets each) that penetrate through the brick cladding at floor level.

Inter-tenancy walls and floors

Although all the walls and floors are created from CLT, different systems have been used to satisfy the various regulations. Standard CLT walls between apartments are clad on each side with a layer of acoustic insulation and two layers of plasterboard, while external walls have the plasterboard fixed directly to the inside of the CLT, with insulation and a drainage cavity layer on the outside face, carefully faced with a final layer of brickwork. Meanwhile inside the lift shaft, the CLT is left exposed, but was fashioned extra thick to allow for charring in the safety factor calculations. Leaving the CLT exposed inside means that no work is necessary to be undertaken in the lift shafts except by the lift installers themselves – a great boon to speediness on site.

Between the floors, fire compartments are separated not just by the CLT panels, but rely on an acoustic isolating layer and a thick layer of screed on top, with a suspended plasterboard ceiling (and acoustic insulating layer) hanging below. The screed acts not just as a fire break and an acoustic deadening layer, but also its mass helps contribute to stop the (relatively) lightweight building from moving in strong winds. Lastly, the screed also acts as thermal storage, capturing heat into the floor and slowly releasing it throughout the night. Waugh Thistleton proudly report that the acoustic measures have helped achieve a further 8dB reduction over that required in Part E of the UK Building Regulations, and that the air-tightness is also two to three times better than required due to the extremely tight tolerances achievable with the use of CLT.

Roof decks

On top of the building, an insulated 'brown roof' system is used to house photovoltaic panels, and the minimal amount of HVAC plant required. All this is hidden behind CLT upstands creating a solid parapet – clad in brick of course, supplanted with a final course of soldier bricks. It is noted as a brown roof rather than a green roof, due to the absence of lush and verdant plant life and instead a more sedate layer of sedum, succulents and pebbles weighing down the drainage membrane.

The last word goes to Kieran Walker, who notes that the proven CLT benefits include not just the dramatically improved programme achievable, no need for a secondary structure, but also the improved precision on site and far greater performance achievable from the building fabric (Walker, 2018). Waugh Thistleton continue to lead the UK in mass timber construction (Waugh Thistleton Architects, 2018). (Figures 18.3–18.11).

Figure 18.3 Dalston Works isometric section.
Source: Courtesy of Waugh Thistleton Architects.

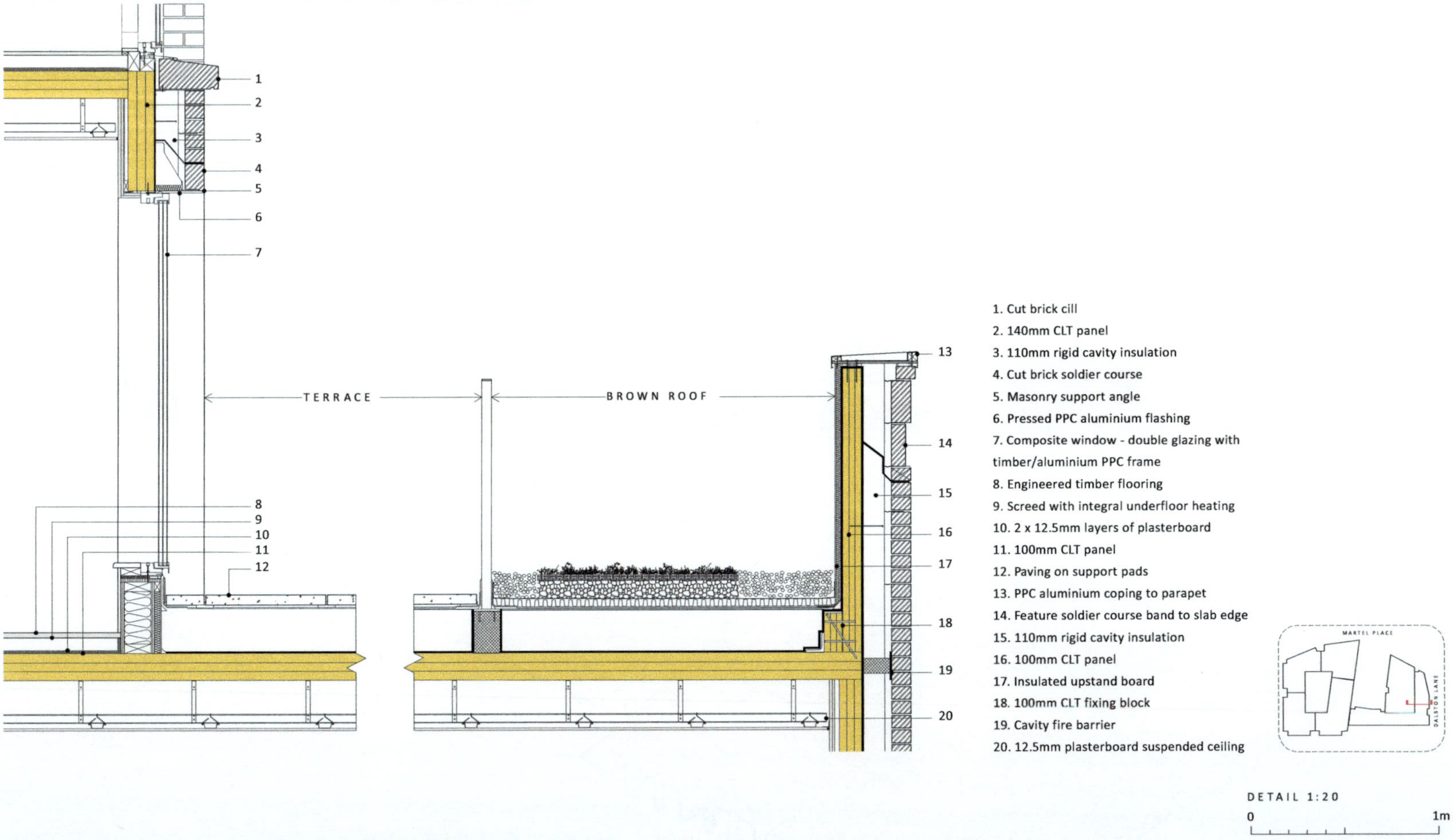

Figure 18.4 Dalston Works balustrade detail.
Source: Courtesy of Waugh Thistleton Architects.

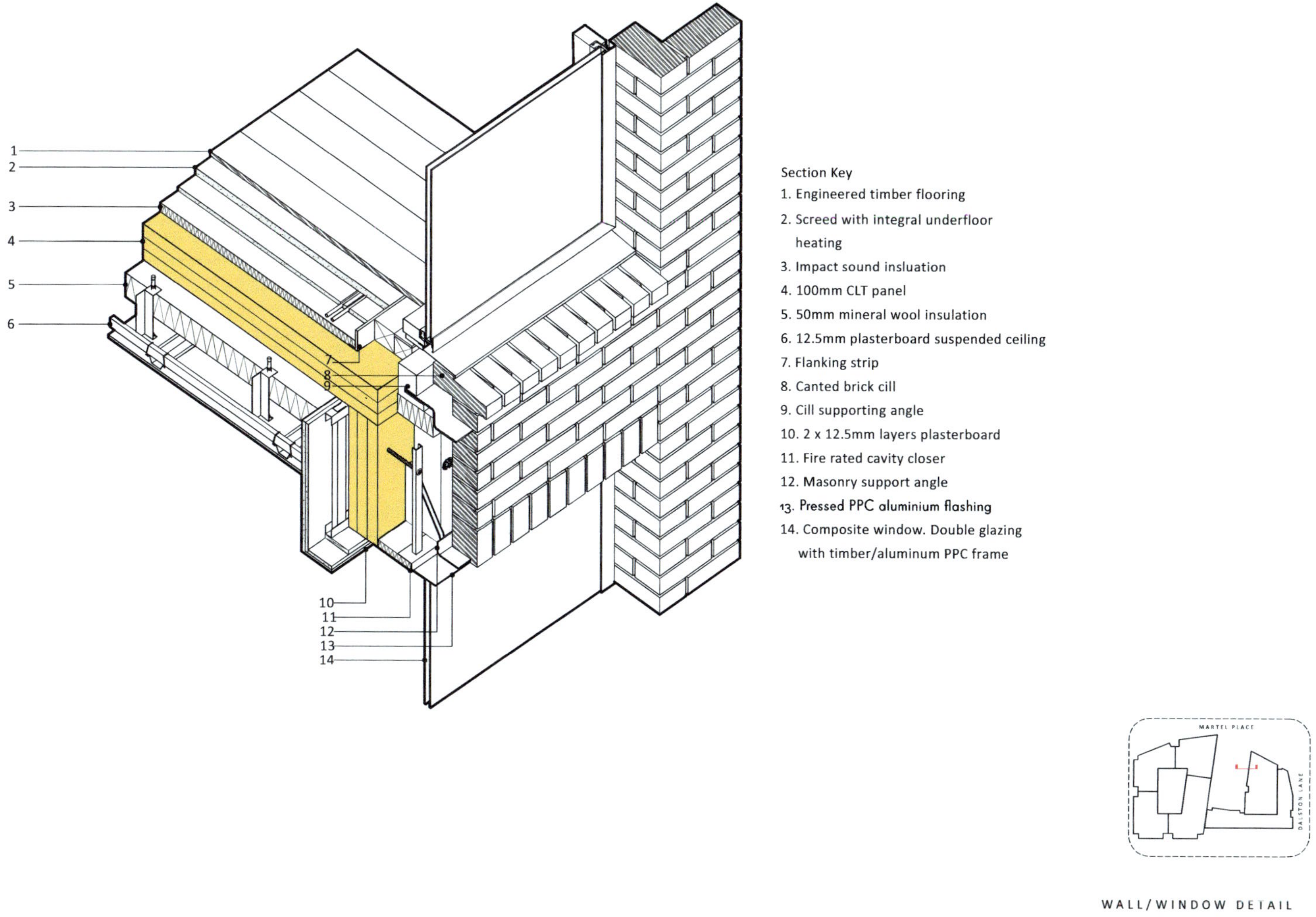

Figure 18.5 Dalston Works 3D external wall/window junction.
Source: Courtesy of Waugh Thistleton Architects.

Figure 18.6 Dalston Works timber walls under construction.
Source: Courtesy of Waugh Thistleton Architects. Photo by Daniel Shearing.

Figure 18.7 Dalston Works brickwork cladding under construction.
Source: Courtesy of Waugh Thistleton Architects. Photo by Daniel Shearing.

Figure 18.8 Dalston Works steel stitch plates connecting CLT to concrete slab.
Source: Courtesy of Waugh Thistleton Architects. Photo by Daniel Shearing.

Figure 18.9 Dalston Works completed scheme.
Source: Courtesy of Waugh Thistleton Architects. Photo by Daniel Shearing.

Figure 18.10 Dalston Works view up to the sky.
Source: Courtesy of Waugh Thistleton Architects. Photo by Daniel Shearing.

Figure 18.11 Dalston Works view with site context.
Source: Courtesy of Waugh Thistleton Architects. Photo by Daniel Shearing.

Works cited

Merrick, Jay. 2017. "High density, low carbon: Dalston Works by Waugh Thistleton." *Architects Journal*, 4 October 2017. Available at: https://www.architectsjournal.co.uk/buildings/high-density-low-carbon-dalston-works-by-waugh-thistleton

Pearson, Andy. 2016. "Dalston Lane: Tall Timber." *Building*, 20 January 2016. Available at: https://www.building.co.uk/dalston-lane-tall-timber/5079749.article

Schuler, Timothy. 2018. "Dalston Works, the Largest CLT Building in the World." *Architect, 9* October 2018. Available at: https://www.architectmagazine.com/technology/detail/dalston-works-the-largest-clt-building-in-the-world_o

Walker, Kieran. 2018. "Towards a construction revolution." Public address to *Timber Design Conference* in Lithuania.

Waugh Thistleton Architects. n.d.. *Dalston Works*. Available at: https://waughthistleton.com/dalston-works/

Waugh Thistleton Architects. 2018. *100 Projects UK CLT*. London: Softwood Lumber Board and Forestry Innovation Investment + Waugh Thistleton Architects.

19

Chapter Nineteen

Case study 4: Mjøstårnet

Case study 4: Mjøstårnet

Øystein Elgsaas

DOI: 10.4324/9781003123873-19

Figure 19.1 Mjøstårnet in the snow, next to heated public baths.
Source: Courtesy of Voll Arkitekter. Photo by Øystein Elgsaas.

Team info

Project title – Mjøstårnet (Figure 19.1)

The name 'Mjøstårnet' is Norwegian and means 'The tower of Lake Mjøsa'

https://www.moelven.com/mjostarnet/

Architects – Voll Arkitekter – lead architect Øystein Elgsaas (www.vollark no)

Engineers – Sweco

Contractors – Hent AS+ Moelven Limtre AS for timber

Client – AB Invest AS

Date of construction – 2017–2019

Project info

Site/address: Nils Amblis Veg, Brumunddal, Norway

Number of storeys – 18 floors total, 85.4m tall. Mixed use

Ground floor – lobby, reception, restaurant, with an additional 4,500m^2 public bath

Level 2 – building services and conference floor

Levels 3–7 – office floors

Levels 8–11 – 72 room hotel, 18 rooms per floor

Levels 12–16 – 33 total residential apartments, with balconies overlooking the lake

Levels 18–19 – 3 residential penthouse units, public viewing terrace and exhibition room

Gross floor area: 10,500 m^2 not including public baths

Net area per floor: 640m^2

Number of apartments: 10 × 1 bed, 15 × 2 bed, 5 × 3 bed units

Approx. construction cost: NOK 450 million – US$79 million

Awards: World's tallest timber building, awarded by the Council on Tall Buildings and Urban Habitat (CTBUH) (Figure 19.2)

Concept

Norway has a long tradition of building with wood, but until 1997 it was forbidden to build wooden houses taller than three floors in Norway. This is due to the old brick enforcement law, which emerged after the city fire in Ålesund in 1904 – the biggest city fire in Norwegian history, which left 10,500 people homeless.

Much has happened since Ålesund broke, and with new knowledge, we are again striving to raise the bar on what is possible, pushing the wood construction technology further, towards a more climate-friendly built environment as a result. We all have a responsibility to find new ways of thinking. And not only thinking, but actually doing.

Figure 19.2 Mjøstårnet typical apartment floor plan.
Source: Courtesy of Voll Arkitekter.

Site context

Brumunddal is a small city with 10,000 residents, 140km north of Oslo, the capital of Norway. The site lays on the riverbank of Brumunda, overlooking Lake Mjøsa, Norway's largest lake. Mjøstårnet is the first time in history that something like this has been constructed on this scale. Together with our client AB Invest and constructor Hent AS, we began to sketch out the project in early 2015. Together with Moelven Limtre, Norway's leading glulam manufacturer and the engineers at Sweco, the team came together, to construct an 18-floor timber building in Brumunddal. Mjøstårnet is a symbol of the 'green shift', and proof that tall buildings can be built using local resources, local suppliers and sustainable timber materials. That is part of our sustainable movement being put into practice.

Programme

The entrance for the tower and the public bath is from a small plaza to the north-west. The tower has a common lobby, used by everyone in the tower. Outdoor space and public gardens are situated to the south, facing the lake and are protected from the noise from the main road. The ground floor is public, with lobby, reception and restaurant. The new building is complemented by a public swimming baths with two 25m length pools in the low-rise building adjacent to the tower.

Timber talk

The surrounding area of Brumunddal is famous for its forestry and wood processing industry. Our client says that standing on top of the viewing platform of Mjøstårnet, you can 'actually see' where the timber comes from and where it was processed.

The statement is not far from the truth.

Norwegian forest owners have long and good traditions for sustainable management of forest resources. When forests are harvested in Norway, forest owners are required to facilitate new forest growth – either by planting or by letting seed trees provide natural growth.

Norway was the first country in the world to map the gap between harvest and growth, already in 1919. In total, Norwegian forests accumulate about 25 million cubic meters of timber every year. Spruce accounts for half this growth. Every year, about 10 million cubic meters of timber are taken out.

After 80 years of growing in the surrounding areas of Brumunddal, reaching the correct height and strength, the spruce is harvested and the timber is cut to standard board size planks at the local saw mill and then processed at Moelven industry, where they make the final glulam products used in the construction. Moelven is just a 15-minute drive from Brumunddal, and the factory responsible for producing the façade elements is situated just across the road.

Mjøstårnet consists of 3,500m^3 of timber or about 14,000 trees. A healthy and sustainable forestry is therefore very important if a project like Mjøstårnet and other timber buildings are to succeed. It was important that the design for the building reflected the 'green shift' in a positive way, as Mjøstårnet is to be a timber building for the future. The budget was about 79 million US dollars for the whole project, including the public bath. The use of timber in the construction needed to compete with a traditional steel and concrete building in cost to be realised, and for this system to be a contender for future building projects. We had to rely on pre-produced elements and repetitive design elements to achieve this.

Construction issues

Early on in the design phase, the overall shape of the building took its base in the most efficient hotel room layout, creating a narrow rectangular shape. The building footprint is approximately 17m × 37m. The roof was slanted, giving it an ideal angle to place solar panels, which also emphasized the environmentally friendly design concept. The tower was also raised from the ground to withstand a 200-year flood.

The shape gave the building a distinct silhouette and it is easy recognisable when you spot it from afar. Our initial idea was to design the tower with an exoskeleton, and just a transparent shell wrapped around on the outside to protect the timber construction from the weather.

The climatic walls would stand on the inside of the construction, because: why build a tower made from wood, if you can't see the wood construction? To further enhance the visibility of the wood construction, we had the tower raised above ground level, separating the tower from the two lower storeys and making the tower rest only on its main glulam columns. In this way, visitors were given the chance to really embrace the construction, just like hugging a tree.

The first draft of the tower created a semi-transparent building with a lot of glass in its façade, with the focus on showing the wood construction to the outside. But the complexity of the construction was too great, and we had to abandon this design. We had to be reasonable, and search for a more solvable construction, simplifying and reducing the elements in play, to make it possible to reach a buildable solution within time and budget.

Instead of glass, we continued our work with the idea of using wood panels as the main façade material. It is a timber house after all.

By working with different-sized boards, and changing the direction in which we placed them, we created a layout which reflected the main construction, and by interlacing the boards we also created a playful pattern. The amount of glass was reduced to suit its function and at designated areas showed the construction. Diagonal beams where placed as much for design purposes as for stability. The end calculation of the construction was not yet ready, so in close collaboration with Moelven and Sweco we kept our options open as long as possible.

To the north, a massive glass wall was implemented to show the elevators' vertical movement, and thus giving the façade towards Brumunddal more life. We also developed a construction system for the slabs, making it possible to have the columns free standing towards the outer walls. In this way, the columns play a more vital role for the interior design of the building as free standing columns, rather than passive columns connected to the façade as normally done. Quite the opposite to the previous concept. But this solution provided an asymmetrical load on the corner columns which was much larger in size than the rest, and could not follow the same system, so once more we had to search for a different solution.

Construction

Before the final stage of the design process, there was a breakthrough in the calculation of the glulam structure. Finally, we landed a system that was both efficient and strong enough to deal with the height, fire proofing and wind forces. The main load bearing consists of large-scale glulam trusses along the façades as well as internal columns and beams. The trusses handle the global forces in horizontal and vertical directions and give the building its necessary stiffness. CLT walls are used for secondary load bearing of three elevators and two staircases. The CLT does not contribute to the building's horizontal stability.

The building is anchored to the bedrock about 40–50m below.

From the second floor up to the eleventh, hollow-box floor elements based on MoelvensTrä8 building system are used as slabs. This system is made of glulam and laminated veneer lumber, with the cavities inside the elements filled with Rockwool. Most elements have a 50mm concrete screed on top.

The floor elements use less timber than solid CLT decks. Moreover, they are light and quick to assemble, and can handle both acoustic and fire requirements. Floor spans of almost 10m can be achieved with this technology, which increases architectural flexibility compared to other timber-based floors. The maximum floor span in Mjøstårnet is 7.5m.

Floors 12–18 are 300mm thick concrete made of a precast element at the bottom and an in-situ layer of topping concrete. Replacing wood with concrete on the upper floors means that the building will be heavier towards the top.

The extra mass is necessary to comply with comfort criteria for wind-induced acceleration in residential buildings. A maximum horizontal deflection at the top of the building of 140mm has been calculated, which is very little for a building of this height.

However, dynamic accelerations are critical for timber high-rises. Peak accelerations due to wind on level 17 are on the limit of what is acceptable for residential buildings. Monitors are placed on Mjøstårnet to learn more about how timber high-rises react to wind, in general.

Each floor acts as a diaphragm that lends stiffness to the timber skeleton. Both the timber and concrete floors are supported by glulam beams, with typical sections up to 625mm × 720mm for the concrete floors. The largest diagonal trusses in the structure have a cross section of 625mm × 990mm. The largest corner columns are 625mm × 1490mm.

All glulam elements are connected by means of slotted-in 8mm steel plates and 12mm dowels. The structural timber elements are placed inside the building envelope, which protects them from rain and sun, thus increasing durability and reducing maintenance.

The envelope consists of large prefabricated façade elements which are attached to the outside of the timber structures. The panels come with insulation and external panels already fixed. Wall elements do not contribute to the global stiffness of the building.

The individual truss members for Mjøstårnet were coded, bundled together and shipped to site, without any trial assembly. Given the complexity of the joinery, the uniqueness of each piece and the precision required, this represented a considerable risk. However, only one out of several hundred glulam elements needed to be re-machined, so the anticipated savings in transportation cost, fabrication and erection time were successfully realised.

The assembly itself was carried out using a large crane on the construction site. No scaffolding was used; four to five floors were lifted at a time. It only took one year to complete the structure.

During construction, the top of the CLT panels and the end grain of glulam columns and diagonal braces were temporarily protected from the weather with plastic. When installation of the exterior wall panels was completed, moisture meters were used to ensure that the wood components had dried sufficiently before being covered or enclosed.

However, the majority of the structure remains exposed to view.

Fire safety

At first people are at bit sceptical when it comes to wood construction and fire safety.

> Wood burns people say – but so does steel, we answer. Of course, if you try to make a fire with a large steel beam, you will struggle, but if you use steel wool instead, it catches fire at once. It is that principle that is being used with glulam or CLT wood. Small wood easily catches fire but glue them together and make the cross section large enough, and you will struggle to make it burn.

In the event of a fire, wood is very predictable, and the fire development is slow. The smoke development in a wooden building is moderate and the tree will be able to maintain both the carrying capacity and the stability of a fire much longer than other materials.

Burnout tests performed in 2016 at SP Firetech in Trondheim, Norway, where large glulam columns were put in a furnace to undergo an ISO-fire for 90 minutes, showed that the glulam continued to char a little more after the burners were shut off. But after several hours, the temperature in all columns tested was declining and the burning stopped. This proves that the large glulam columns will self-extinguish and prevent a building from collapse. This provides better escape time for those in the building.

Many people will recognise this themselves after placing a heavy cube of wood in the fireplace. It will not burn up if you do not add the small kindling wood. The big cube will soon get a layer of charcoal, but the fire will eventually die out on its own.

The fire strategy report for this project states that the main load bearing system must be designed to withstand 120 minutes of fire. Secondary load bearing, such as floors, must withstand 90 minutes of fire. The fire resistance can be obtained by calculating the remaining cross section after charring according to Eurocode. The required two-hour fire resistance is achieved by oversizing the glulam elements by approximately 80mm in each dimension.

When it comes to robustness, the structure is designed to sustain the loss of the horizontal stiffness of one timber floor. It can also carry the impact load of a timber deck falling down on the floor below.

Mjøstårnet is designed to withstand a complete fire. This means a fire that is allowed to develop freely without active firefighting effort. Two separate sprinkler systems are installed. Each floor, hotel room, offices and apartments are constructed as separate fire cells. There is a fire control room on the ground floor with full overview of the building.

Visible wood in escape routes as well as internal walls in the main staircase and elevators have fire retardant painting. Exposed walls in the escape staircase are covered with plasterboard.

There are Firestop barriers in the façade to prevent fire from spreading upwards. Steel plates and dowels in the connections are embedded deep into the timber (85+ mm). Gaps and slots between beams, columns and plates are fitted with an expanding fire strip. This material expands about 20 times when the temperature reaches 150°C, protecting the steel from reaching critical temperatures.

As mentioned, the façade is composed of prefabricated elements. Fire requirements apply not only to the surface of the cladding, but also to the back of the woodwork and the underlying construction. Non-combustible insulation is also used. Cavities in the façade are broken at each floor, to prevent fire spreading in the cavity behind.

The woodify product used as cladding, is a thermo-wood product, made from Swedish pine that has been modified through a heat and steam process, making it non-toxic, dimensionally stable, resistant to decay and resin free.

The fire impregnation is achieved through a carefully monitored vacuum and pressure process. After the materials are subjected to vacuum, the fire protection fluid is applied under pressure, and after drying, the fire protection remains within the cell structure of the woodwork.

The fire protection lasts the life of the material. Since the protection works within the cell structure of the woodwork, fire-resistant wood can therefore be adjusted on site.

Each type of tree has unique properties and reacts very differently to the impregnation process. It is therefore very important that each tree species undergoes an individually developed impregnation process, as well as undergo individual fire tests. The fire measures carried out at the Mjøstårnet make this one of the safest buildings along Mjøsa.

Final design

The facade elements were produced off site, and because of transport regulation (even if the factory only was 300m down the road), the best way of utilising the pre-produced elements was to place them horizontally for each floor. This made the most of the length and height of each element, thereby reducing the number of elements needed and the amount of time needed to install them.

Because of the horizontal joints, we had to implement a visible flashing for each floor. The flashing also works as a shield in case of fire inside the void of the construction, preventing fire from spreading.

We chose vertical board panels with extra vertical tracks, helping to control drips from any surface and lead them straight down. All the boards are one storey high with no joints.

Square and thicker planks were used to enhance the verticality in the project by placing them in a rhythmic pattern on the outside of the main panels, giving the façade more texture and depth.

To keep some of the original ideas with board panels in different directions, horizontal boards were placed below and in between windows at designated areas, and through different stages of the design process these areas were the subject of many discussions. Should these areas have colour, should these areas have another material than wood or could we perhaps use larger windows instead, to create a less massive expression of the façade and give more light and better views from the inside. In some areas the large trusses take up a lot of the glass area, but in the end, the client chose this solution.

Instead of raising the tower from the ground to withstand potential flooding, the overall terrain level was raised instead, creating a secure evacuation route for the inhabitants and a way in for emergency personnel.

Due to heavy snowing during winter months, the angled roof could potentially create large snow slides. So, instead of exposing the timber structure on the ground level, we chose to create an open roof structure, replicating the structure of the tower. Situated in a more controlled environment, we could safely showcase the structure as initially intended with minimal protection and still keep our original silhouette. The risk of snow slides were also minimised, and at the same time gain a more characteristic expression. Unfortunately, this meant a reduction in the number of solar panels but we kept some on the south façade of the tower and the public baths.

For the timber floors, plumbing and drainage pipes had to be coordinated with the construction of the floor slabs, as penetrations for services could potentially come into conflict with the glulam structure and potentially weaken it.

The plan layout is similar on each floor which has the same function, for practical and economic reasons, so a wide spread of different windows and window placement was not possible. Similarly with the prefabricated wall elements, to keep the cost down, they had to be as alike as possible. We also had to make sure that the diagonal trusses did not steal too much of the daylight when it came in conflict with the window placement and that every resident had access to their balcony without needing to crawl underneath or climb over a glulam beam.

From the outside, you can get still get a glimpse of the glulam trusses stretching from the ground floor up to the top, through the windows of the building. This helps create a readable façade where you can see and understand the wood structure as originally intended. The glulam construction on the roof is, of course, also a help in reading the wood structure of the tower.

The idea of the glass wall to the north was kept, but done by using several smaller windows placed in three vertical ribbons instead, one for each elevator. This kept the idea of seeing the elevators from the outside and giving the user a view of the construction of the tower and its surroundings as they take the elevator. Having tested it for myself, I am glad to say that this is one elevator ride you will remember: the view towards Brumunddal is amazing.

The large columns and trusses had, of course, an impact on the plan layout, and we had to integrate the columns as part of the architecture itself more than we usually would to provide a satisfactory solution to the layout. In the end, many of the columns, which have a high, furniture-level degree of finish and also work as a nice interior design element.

In Norway, elements of wood in the indoor environment are very common. The visible grain in the tree gives a positive connection to nature. Several studies show a lower level of stress in rooms with visible wood interiors. Physically, there are also effects that give positive results with wood interiors. The low heat conductivity of wood gives greater tolerance to air temperatures. In addition, the ability wood has to absorb and emit moisture has a positive contribution to the energy balance in the house.

The apartments were sold quickly, in fact, most of the buyers asked to know in which apartments the columns and beams were visible, because those were the apartments they wanted to buy. This shows that the market has accepted that mass wood structures are as safe as their traditional steel and concrete equivalents.

Even if the tower looks very different now, compared with the first sketches, it is easy to see that some of the original ideas have survived. And if you are ever passing by on the highway, you will be amazed when you realise that this building is made from wood.

We think it's exciting to see that wood construction has gained a new renaissance and we are proud to be able to help develop wood architecture to new heights. Mjøstårnet is not the blueprint of a tall timber building, but a contributor to further sustainable development.

Different materials will be used side by side for many years to come. Sustainability-wise, the most important aspect of this building is to show that it is possible to build large, complex timber buildings, and in so doing, inspire others to do the same (Figures 19.3–19.10).

Figure 19.3 Mjøstårnet typical hotel floor plan.
Source: Courtesy of Voll Arkitekter.

Figure 19.4 Mjøstårnet restaurant interior.
Source: Courtesy of Voll Arkitekter. Photo by Øystein Elgsaas.

Figure 19.5 Mjøstårnet interior showing large timber members exposed.
Source: Courtesy of Voll Arkitekter. Photo by Øystein Elgsaas.

Figure 19.6 Mjøstårnet timber staircase.
Source: Courtesy of Voll Arkitekter. Photo by Øystein Elgsaas.

Figure 19.7 Mjøstårnet hotel bedroom.
Source: Courtesy of Voll Arkitekter. Photo by Øystein Elgsaas.

Figure 19.8 Mjøstårnet ground floor lobby.
Source: Courtesy of Voll Arkitekter. Photo by Øystein Elgsaas.

Figure 19.9 Mjøstårnet exterior timber cladding.
Source: Courtesy of Voll Arkitekter. Photo by Øystein Elgsaas.

Figure 19.10 Mjøstårnet in context with the landscape.
Source: Courtesy of Voll Arkitekter. Photo by Ricardo Foto AS.

20

Chapter Twenty

Case study 5:
Brock Common

Case study 5: Brock Commons: Tallwood House

Russell Acton and Guy Marriage

DOI: 10.4324/9781003123873-20

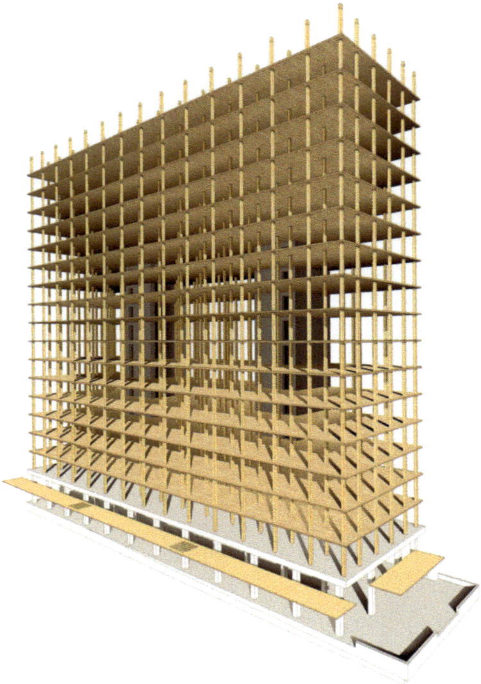

Figure 20.1 Brock Commons Tallwood House mass timber structure.
Source: Courtesy of Acton Ostry Architects.

Team info

Project title – Brock Commons Tallwood House (Figure 20.1)

https://www.actonostry.ca/project/brock-commons-tallwood-house/

Architects – Acton Ostry Architects

Russell Acton, Principal Design Architect; Mark Ostry, Principal Managing Architect

Matthew Wood (Project Lead), Rafael Santa Ana, Andrew Weyrauch, Gjergj Hondro, Nebojsa Slijepcevic, Nathaniel Straathof, Warren Schmidt

Engineers:

Tall Wood Advisor – Architekten Hermann Kaufmann ZT GmbH (Schwarzach)

Hermann Kauffman, Christoph Duenser

Structural – Fast + Epp

Paul Fast, Bernhard Gafner, Robert Jackson, Jimmy Li

Building Code & Fire Science – GHL Consultants Ltd

Andrew Harmsworth, K., M. Gary Chen

Building Envelope – RDH Building Science Inc

Brian Hubbs, Graham Finch, Robert Orlowski, Matthew Benson

Mechanical, Electrical & LEED – Stantec, Inc

Michael Dhont, Svetlana Vujic, Jim Jay, Michael Fletcher, Graham Twyford-Miles, Sean Moore

Acoustic – RWDI Air, Inc

Daniel Lyzun

Energy Modeling – EnerSys Analytics, Inc

Curt Hepting

3D Modeling & Virtual Construction – CadMakers Inc

Javier Glatt, Nicolas Cantin

Construction Manager – Urban One Builders

Brent Ohlund

Wood Structure Erection Consultant – Seagate Structures Ltd

Ralph Austin, Nathan Bergen

Concrete Formwork Consultant – WhiteWater Concrete Inc

Kyle Smith

Prefabricated Facade – Centura Building Systems Ltd

George Beja, Ash Botros

Landscape – Hapa Collaborative

Joseph Fry, Hanako Amaya

Civil – Kamps Engineering Limited (Toronto)

Michael Kamps

Project Manager – UBC Properties Trust

Rob Brown, David English

Client – University of British Columbia

John Metras, Andrew Parr, David Kiloh

Date of construction – started November 2015 – completed May 2017

Time needed for the timber construction = 2 months

Time needed for the whole construction = 17 months

Project info

Site/Address: 6088 Walter Gage Road, UBC, Vancouver, BC

Number of storeys – 18 storeys. No basement, ground floor social/study space, levels 2–17 student residential, level 18 residential and student lounge

Awards received

2018 International Prize for Wood Architecture

2018 Lieutenant Governor of BC Award

2018 Sustainable Architecture & Building Green Award

2018 Canadian Wood Council Wood Works Architect Award

2018 Canadian Wood Council Wood Works Innovation Award

2018 Canadian Wood Council Wood Works Engineer Award

2018 Council on Tall Buildings & Urban Habitat Award Finalist

2018 Lieutenant Governor of BC Award for Engineering Excellence

2018 Vancouver Regional Construction Association Gold Award

2017 Premier's Award Promoting Innovation and Excellence

2017 Canadian Wood Council Special Jury Award

2017 Construction Dive Five Favorite Projects of the Year

2017 Institution of Structural Engineers Innovation Award

2017 NCSEA Excellence in Structural Engineering Award

2017 Fast Company Innovation by Design Finalist

Site coverage: Gross site area 2,314m^2 (24,908 sq ft)

Gross floor area: 15,120m^2 (162,750 sq ft)

Typical residential floors: 824m^2 (8,870 sq ft)

406.4 m^2 × 16 studio dwelling units

230.4 m^2 × 2 quad dwelling units

28.8 m^2 communications/housekeeping

14.4 m^2 mechanical

3.6 m^2 electrical

140.6 m^2 circulation

Number of apartments: 272 studio rooms, 33 quad bed units: 404 beds in total

Average size of apartments: Studios: 25.4m^2 each (273 sq ft each), quads 115.2m^2 each (1,240 sq ft each)

Construction budget: CAN $40.5m ($2,668/m^2) ($248/sq ft in 2017) (Figure 20.2)

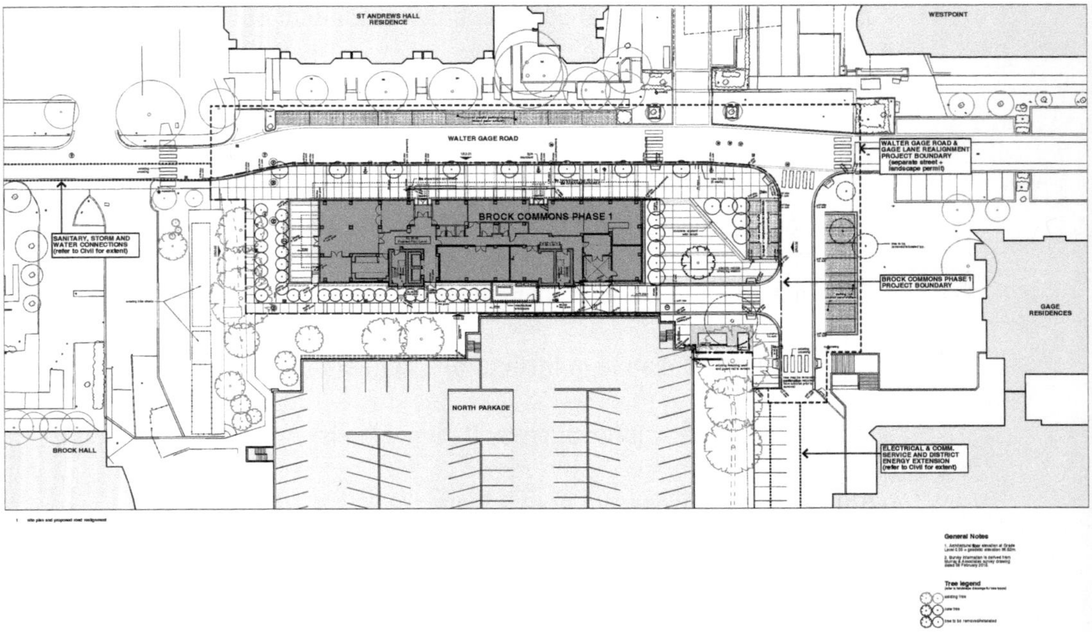

Figure 20.2 Brock Commons location plan.
Source: Courtesy of Acton Ostry Architects.

Concept

Brock Commons Tallwood House, is an innovative LEED Gold certified, 18-storey, 53 m high, tall wood student residence located on the University of British Columbia Point Grey campus, which at the time of completion in 2017 was the tallest contemporary mass timber building in the world. The project was part of the 2013 National Resources Canada Tall Wood Building Demonstration Initiative that aimed to demonstrate that mass wood structures are a viable structural option for high-rise buildings.

Tallwood House is a model for a future that features extraordinarily ordinary mass wood buildings that are quick, clean and cost effective to construct, and that maximise carbon sequestering and reduction of greenhouse gas emissions.

A key mandate set by the university was that the project had to meet the construction budget and demonstrate to the development industry that tall mass timber structures are an economical and viable alternative to concrete structures. The project team, comprised of design professionals, fabricators, and erection and construction management personnel, rallied around a common goal of simplicity and efficiency of design, construction, budget and schedule. The key to receiving approvals and achieving economic viability was the 'keep it simple' design approach that makes the building appear deceptively ordinary, due to the encapsulation of the wood structure with fire-rated gypsum board to achieve required fire protection ratings.

Mass wood construction is in its infancy. For the industry to grow and evolve, tall timber buildings must become commonplace and ubiquitous, which will only happen if they are affordable to design and build. To be truly environmentally meaningful, mass wood structures must be incorporated into buildings of all types and sizes, from the audacious to the everyday, whether the wood structure is exposed or not.

Brock Commons Tallwood House demonstrates that an encapsulated mass wood tower can and will be comparable in cost to a concrete high-rise. Price parity is within reach. The continued evolution of tall wood buildings is the foundation upon which mass wood will make a meaningful contribution to the future sustainability of cities.

Principal means of construction

Level 1 – Concrete transfer structure and two concrete core structures

The building is supported on a reinforced concrete raft slab foundation with concrete columns that support a transfer slab at level 2. The Site Specific Regulation dictated that the ground floor and two stair/elevator cores were to be concrete construction, with the structural design required to follow the not-yet-adopted National Building Code of Canada 2015 that included seismic acceleration values approximately 50% higher than the current building code. Twin 18-storey concrete cores house the exit stairs and elevators, completed in a traditional manner with slip-formed cores by Whitewater (formwork), LMS (reinforcing) and LeFarge (concrete). From ground level outside, there is little to indicate that this building is not concrete all the way up.

Levels 2–18 – Mass-wood structure

Above level 2 however, apart from the cores, the structure is entirely timber. The concrete transfer slab supports 17 levels of mass timber superstructure comprised of five-ply, 169mm thick, CLT panels. The hybrid mass timber structure is significantly lighter in weight than a comparable fully concrete structure. The lower mass and inertia result in reduced resistance to overturning forces during a seismic event, which is counterbalanced by the concrete cores. Lateral loads are carried through the CLT floor plate diaphragms and transferred to the concrete cores through steel drag straps fixed to the surface of the CLT. Columns are set out on a 2.85m × 4.0m grid, and in a daring structural solution the use of a fully two-way CLT slab diaphragm completely eliminated the need for load-carrying beams.

To avoid a vertical load transfer through the CLT panels, a steel connector allows for a direct load transfer between the columns and provides a bearing surface for the CLT panels (Kaufmann et al, 2018). There are 29 CLT panels per floor (464 panels in total), with 78 glulam columns per floor (1,302 columns in total). The columns are created from British Columbia's extensive forests of Douglas Fir, laminated by Structurlam from parallel laminations as Parallam (PSL), supplied by Weyerhaeuser and installed by Seagate Structures. Each column is just 265mm × 265mm glulam reducing down to just 265mm × 215mm above the ninth floor. The highly innovative timber structure is extraordinary for the speed at which its glue laminated columns, cross laminated floor panels, and prefabricated facade were erected in just 66 days, at a rate of two floors per week. This was nine weeks faster than a comparable concrete structure, but without a comparable building envelope enclosure included as well (Parker, 2018).

A cross laminated timber canopy runs the length of the curtain wall base, revealing warm wood finishes at ground floor amenity spaces. Glulam columns are left exposed at an upper level amenity lounge. Hallways feature wood doors and rich umber and ochre accents. The living units are spare and simple with bright white finishes and warmly-hued finishes. The quintessential west coast ocean and mountain views are stunningly spectacular.

Cladding system

A prefabricated façade was installed in conjunction with the erection of the timber structure in order to enclose the building and protect the mass wood from wind-driven rain. The innovative prefabricated façade features corner window-wraps and high-pressure laminate (HPL) cladding that contains 70% wood fibre. The prefabricated façade (there are 22 prefabricated façade panels per floor) was manufactured by Centura Building Systems to stringent tolerances and was comprised of:

- steel stud framing with exterior grade gypsum board (DensGlass sheathing)
- liquid-silicon air and water membrane
- fiberglass thermal clips and steel girts
- semi-rigid stone-wool insulation (ROXUL Cavity Rock insulation on Cascadia clips)

– high-pressure laminate cladding (Trespa panel) made with 70% wood fibre and thermosetting resins.

With the prefabricated panels constructed on a light gauge steel stud frame, they were able to be large – 8m wide by 2.65m high – and craned directly into place. Each panel would form the entire wall of two studio bedrooms, complete with two double-glazed windows (Phoenix Glass) with thermally broken aluminium frames, installed in the factory. The Trespa panels feature a printed wood effect to the external face, indicating the true timber hidden behind the façade. Following installation of the façade panels on site, backer rods, caulking, additional fibreglass batt insulation, poly vapour barrier, and interior gypsum board completed the building envelope.

Construction/design approach to fire safety and seismic issues

Brock Commons Tallwood House was designed and constructed in accordance with the 2012 British Columbia Building Code under a Site Specific Regulation issued by the Province of British Columbia Building Safety Standards Branch, which set requirements for the building to meet or exceed the performance of a comparable non-combustible building.

To facilitate and expedite the Site Specific Regulation approval process, a conservative approach was used to demonstrate fire protection equivalency. The primary fire protection approach was encapsulation of the mass timber structure with fire-rated gypsum board to provide a two-hour fire-resistance rating. With no exposed wood, a more economical grade of wood could be specified, which reduced costs. In addition, the cross-section of the wood components could be reduced in size since there was no need to accommodate charring to achieve fire protection requirements, which further reduced costs. Fire protection measures include an automatic sprinkler system, an emergency back-up water supply tank, and a fire pump.

The approval process included structural peer reviews by two third-party engineers and design reviews by two expert panels. The expert panels, one with structural oversight and one with fire safety oversight, were comprised of over 20 architects, engineers, fire and code officials, building scientists, advanced wood construction research organisations, and UBC engineering and building science faculty.

Approach to sustainability/prefabrication/on the project

Brock Commons Tallwood House targets LEED v4 Gold certification, conforms to ASHRAE 90.1–2010 and has a building energy target set by the UBC Sustainability Office of 135 kWh/m²/year. The building is connected to the UBC District Energy System (DES), which supplies heating for the service water heating and ground floor amenity spaces. Heating for the resident units is provided by electric baseboard. The energy supplied by the DES accounts for 63% of site energy use with 30% of the heat coming from the Bioenergy Research Demonstration Facility (BRDF). The BRDF facility provides 5% of campus electrical needs, which means 5% of the electricity used is from a renewable source.

The high attention to the life cycle of Tallwood House resulted in a building with a holistically low carbon footprint, both on the embodied as well as on the operational life cycle stages. When using ATHENA to perform a 60-year cradle-to-grave Life Cycle Analysis, the project demonstrates an exceptionally low embodied carbon intensity of 236 kg CO_2-eq/m^2 Global Warming Potential (GWP). Compared to a similar, conventional concrete structure, Tallwood House achieved 43% savings on GWP, 60% on Ozone Depletion Potential, 47% on Eutrophication Potential and 35% on all other environmental indicators, achieving exemplary performance on LEED v4 Materials and Resources Credit 1.

The wood used at Tallwood House was sourced from B.C.'s sustainably managed forests, where less than one-third of 1% of public forests are harvested annually, with a legal requirement to regenerate all harvested areas. The timber used stores an impressive 1,753 tonnes of carbon dioxide and avoided production of 679 tonnes of greenhouse gas emissions. The 2,500m^3 volume of wood products used is grown in Canadian and U.S. forests in six minutes and is equivalent to removing 511 automobiles off the road for a year.

Construction cost

One of the biggest challenges facing the tall wood movement is the issue of cost of construction. While many extol the low cost of timber construction, other than Brock Commons, there is little reliable information regarding the cost of encapsulated mass wood and virtually none for exposed mass wood.

At the time of completion in 2017, the construction cost for Tallwood House was CAN $40.5 million or $2,668/m^2 ($248/sq ft). Analysis showed that a comparable conventional concrete building would cost $37.8 million or $2,496/m^2 ($232/sq ft). Therefore, the cost premium for the first-of-its-kind tall wood building was estimated to be approximately 7% of a comparable concrete high-rise.

The mass timber structure for Tallwood House cost approximately $548/m^2 ($51/sq ft), while a comparable concrete structure cost about $516/m^2 ($48/sq ft). Therefore, the cost of the mass wood structural components was approximately 20% of the total construction cost – the same as a comparable concrete structure.

Future tall wood projects will benefit from the experiences gained with the design and construction of Tallwood House and as lessons learned are shared with the industry. By successfully coming in on budget and on schedule, Tallwood House demonstrates that timber high-rise buildings can be economically competitive with concrete buildings. It is anticipated that the 7% first-time cost premium will be reduced and ultimately eliminated as the industry matures, builders gain experience, and construction methods for tall wood buildings are further optimised. As the costs of concrete construction keeps rising, it is expected that future tall wood buildings will cost less and be quicker, cleaner, and more sustainable to build than their concrete counterparts (Figures 20.3–20.13).

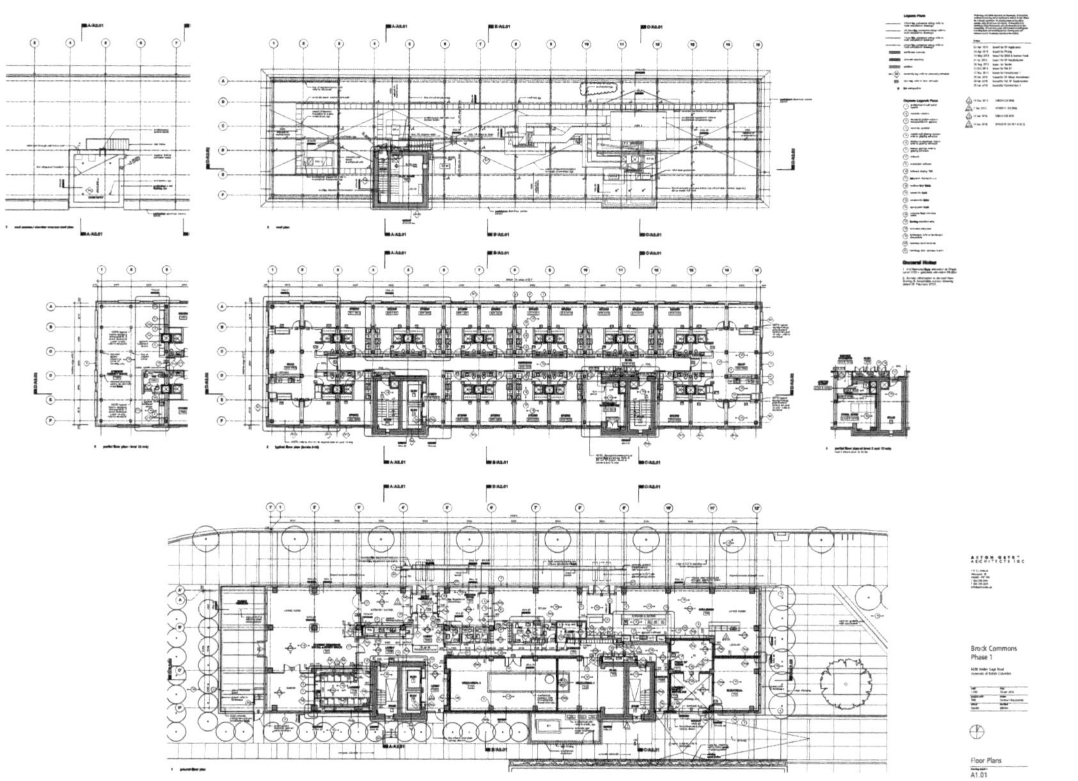

Figure 20.3 Brock Commons student housing typical floor plan.
Source: Courtesy of Acton Ostry Architects.

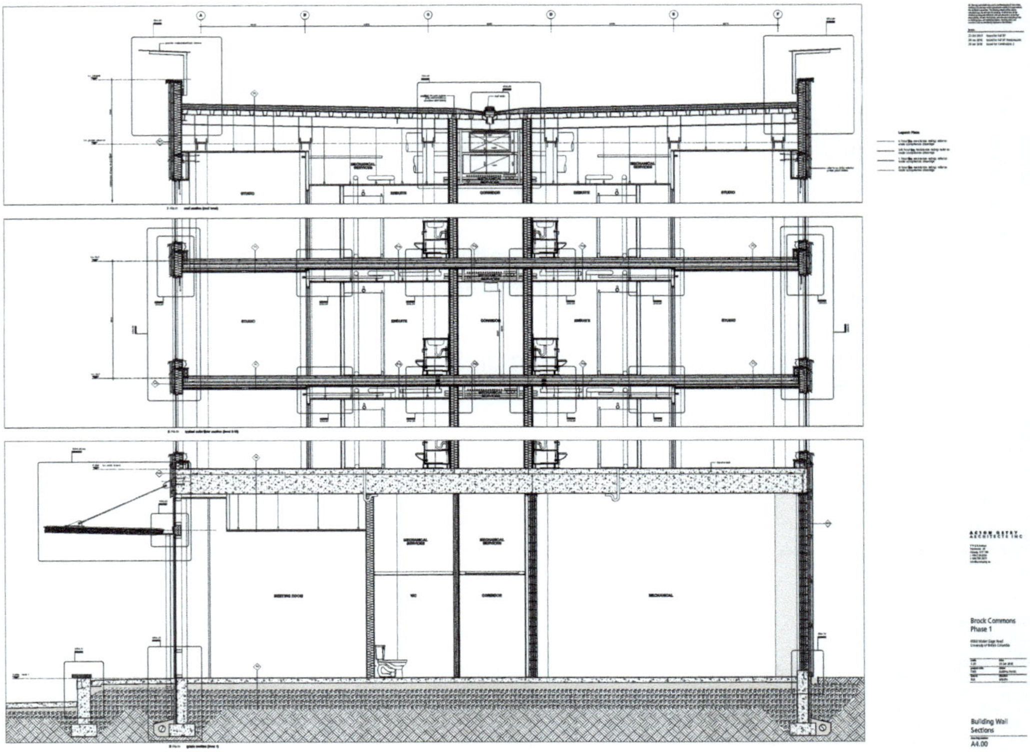

Figure 20.4 Brock Commons cross-section.
Source: Courtesy of Acton Ostry Architects.

Figure 20.5 Brock Commons Tallwood House – prefabricated panel installation.
Source: Courtesy of Acton Ostry Architects.

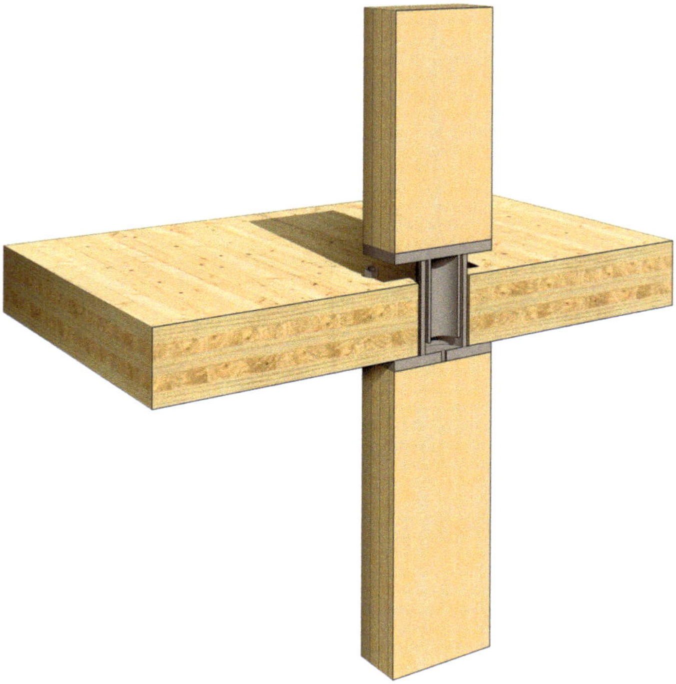

Figure 20.6 Brock Commons Tallwood House – CLT to glulam connection via steel boot.
Source: Courtesy of Acton Ostry Architects.

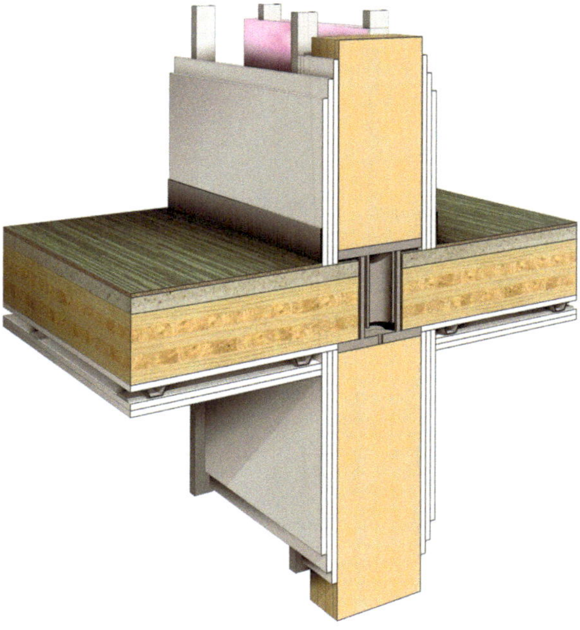

Figure 20.7 Brock Commons Tallwood House – completed connection detail.
Source: Courtesy of Acton Ostry Architects.

Figure 20.8 Brock Commons Tallwood House – week 1 construction.
Source: Courtesy of Acton Ostry Architects. Photo by Pollox Chung.

Figure 20.9 Brock Commons Tallwood House – week 4 construction.
Source: Courtesy of Acton Ostry Architects. Photo by Pollox Chung.

Figure 20.10 Brock Commons Tallwood House – week 8 construction.
Source: Courtesy of Acton Ostry Architects. Photo by Pollox Chung.

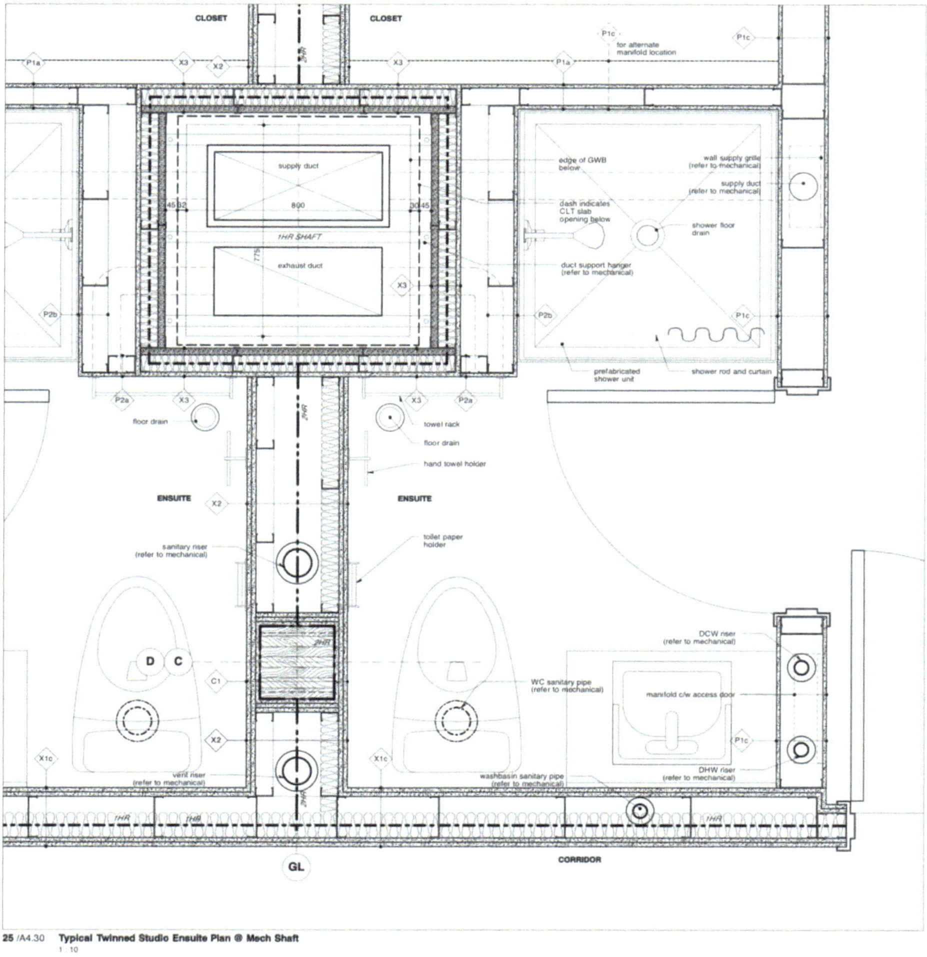

Figure 20.11 Brock Commons – studio apartment wall at bathrooms.
Source: Courtesy of Acton Ostry Architects.

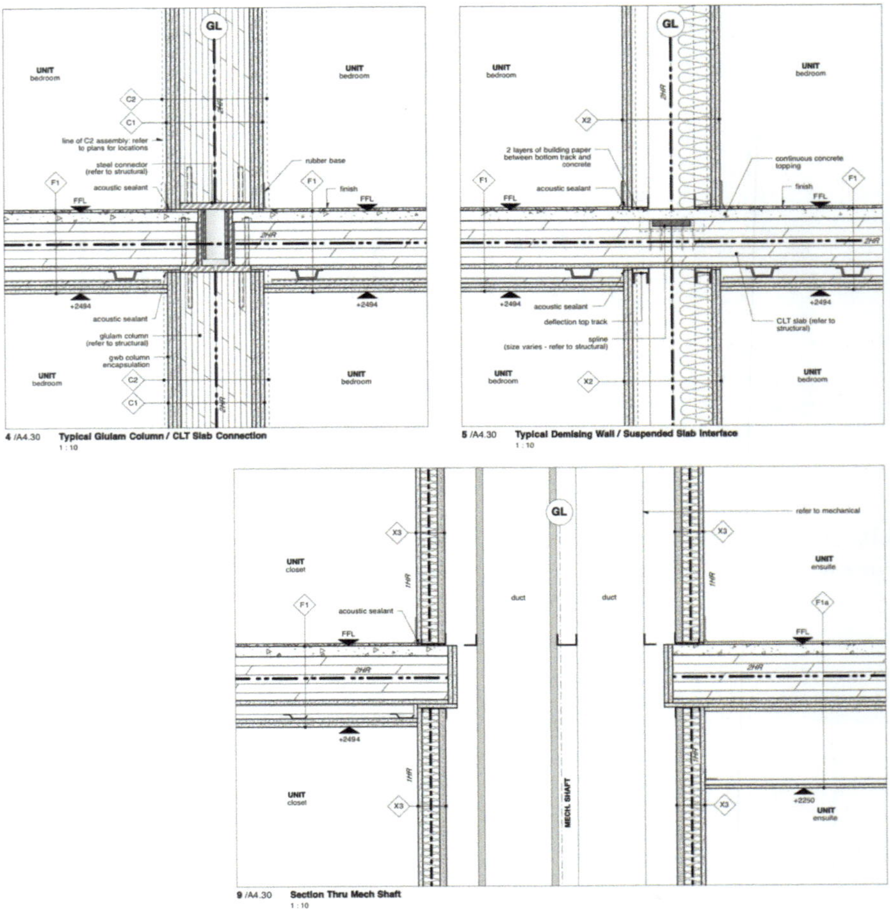

Figure 20.12 Brock Commons Tallwood House demising wall details at service riser.
Source: Courtesy of Acton Ostry Architects.

Figure 20.13 Brock Commons Tallwood House installing panels.
Source: Courtesy of Acton Ostry Architects. Photo by Pollox Chung.

Works cited

Acton Ostry Architects & Fast + Epp: Brock Commons, Tallest Modern Timber Hi-Rise. Available at: https://clean50.com/projects/acton-ostry-architects-fast-epp-brock-commons-tallest-modern-timber-hi-rise/

Kaufmann, Herman, Krotsch, Stefan, and Winter, Stefan. 2018. *Manual of Multi-Storey Timber Construction*. Munich: Detail.

Parker, James. 2018. "Mass appeal." Architect's Data File. Available at: https://www.architectsdatafile.co.uk/news/mass-appeal-brock-commons-tallwood-house-vancouver-canada/

21

Chapter Twenty One

Case study 6: Regensbergstrasse

Joe Kaps and Guy Marriage

Case study 6: Regensbergstrass

DOI: 10.4324/9781003123873-21

Figure 21.1 Regensbergstrasse new multi-storey apartment building.
Source: Courtesy of Dreicon.

Team info

Project title – Ersatz Neubau Mehrfamilienhaus Regensbergstrasse (New multi-storey apartment building in Regensberg Street) (Figure 21.1)

Architects – Dreicon AG – architects Joe Kaps, David Jiranek, Martin Sperber, Artemis Papandreou, Noelis Carangelo

Engineers (Structural and Building Physics) – Adali Engineering GmbH

Engineers (Mechanical, Plumbing) – Fortunato Engineering AG

Engineers (Electrical) – Bühlman Engineering AG

Fire Engineering – Brandschutzpartner

Landscape Architect – Stefan Rotzler

Planner – Dreicon

Client – Implenia Vorsorge

Contractors – Implenia Schweiz AG

Date of construction – 2021

Project info

Location – Regensbergstrasse 30, CH-8050 Zurich, Switzerland

Number of storeys – five storeys residential plus basement

Site area – 1,474 m^2

Building footprint – 520m^2 + 60m^2 terraces

Built area – 2,500m^2 + 885m^2 basement

Net Living area – 1,900m^2

Number of apartments – 23 apartments total: 9 @ two-bed, 8 @ three-bed, 3 @ four-bed, 3 @ five-bed

Average size of apartments – two-bed = 62m^2, three-bed = 82m^2, four-bed = 106m^2, five-bed = 126m^2

Approximate construction cost. 8.8 million Swiss francs (Figure 21.2)

Concept

An existing apartment building of 21 units was to be demolished due to its age and condition, unable to continue to meet modern building codes. An initial proposal was made that would enable 22 units on site, but when Dreicon got involved, they won the limited competition by careful and strictly regimented planning, permitting the developer to create a total of 23 new apartments. Housing in Zurich is taken very seriously, with Swiss pride in the precision and good nature of the development.

A simple building diagram based on a simple structural and MEP distribution concept makes for a much rationalised scheme, which together with an optimised use of the complicated plot setbacks and building length limitations allowed for an improved lettable area and indeed number of apartments.

Context

Zurich sits on the edge of the Zurichsee, a long lake at the base of the Swiss Alps. It has a climate that gets stifling hot in summer and yet cold

Figure 21.2 Regensbergstrasse location plan.
Source: Courtesy of Dreicon.

and snowy on occasions in winter. Zurich is a wealthy city, the home of the Swiss banks, and therefore has high standards for housing its workers. Hemmed in by low-rising hills, and with an excellent and elaborate system of trams, most people in Zurich live in apartments and use public transport extensively. Apartment building is a vibrant activity in this city and building standards are high.

In the neighbourhood of Oerlikon, just a short distance from central Zurich, Regensbergstrasse is close to Waldgarten tram stop on the number 7 tram line into the city. The Zürichberg forest is nearby; dense and heavily-wooded with much natural wildlife. Although Regensbergstrasse is a built-up area with many multi-storey apartment buildings, this site looks out onto a small neighbourhood park across the street. While the building has a definite air of quality, the architects insist that this is not an expensive, upmarket project. Instead, the level of quality achieved here is just the same, good, basic standard of quality that is expected throughout medium density housing developments in Switzerland.

Planning

A single central core, with one lift and a single open stairwell, is roofed over with a double-glazed skylight, throwing sunlight down into the very heart of the building. A small racetrack of gallery access to the apartments keeps circulation to a minimum. Apartment planning here is focused and uncomplicated. All bedrooms have an external window, some also with a window onto the deck/balcony. Kitchens are either a galley or a simple L-shape, not extravagant: all about 4.5m long. Bathrooms are allocated generally at one per apartment, sometimes two for the larger units, and apartment sizes are small but not tight – a mixture of one bedroom, two bedroom and three bedroom. Living rooms are adequate but not overly large, for these are not large apartments. Only the two penthouse apartments are noticeably bigger and more luxurious.

A further planning feature of the Swiss system is that space for storage is strictly enforced – it is simply not something that can be left out. There is a requirement for a certain amount of storage upstairs in each apartment, and further storage is made available downstairs in the basement – a logical requirement in a city where almost everyone skis for several months a year. Despite the closeness of the number 7 tram, there is also a requirement for one car park space per apartment as well.

Construction information

The building is constructed from concrete – columns, slabs, lift shaft and shear walls. There is a simple and yet very strong form to the design, with six shear walls radiating out like a windmill from the central lift and staircase. Corridors have been cut short and reduced to an absolute minimum in the design, one of the factors that allowed Dreicon to gain enough space for two extra apartments. The concrete slab is 250mm thick, which also includes multiple conduits for wiring to run through, with layers of steel reinforcing above and below. On top of the slab sits an acoustic barrier, and then a fire resisting barrier, topped with a DPC. On top of the DPC sits a 50mm screed with inset heating coils, and finally on top of that is the final timber floor covering. The Swiss-based architects are, of course, highly attentive to detail. Overall floor thickness therefore is over 300mm, but noise transmission is very much blocked.

Apart from the 250mm thick in situ concrete shear walls, other minor walls are concrete blockwork 150mm thick, but these are also carefully constructed to ensure a seamless acoustic experience for the residents. All walls containing services are constructed separately from the inter-tenancy walls, so that no services-related noises can break the acoustic envelope. External walls are also concrete block, overclad with a minimum of 100mm insulation and then completed with a quality rendered plaster finish. Dreicon have ensured that while the main structural walls are finished in a green colour to the render, the precast balcony panels are left in a natural state and simply painted a lighter colour.

Glazing is triple-glazed as standard, with window frames made from PVC with an external anodised aluminium protective layer coloured to tie in with the plaster render. Heat loss through the windows is very low as a result, allowing the apartment to be warm and comfortable inside while the snow settles deeply outside.

Sustainability

Switzerland has strongly controlled building standards, which include a need for very thorough thermal and acoustic insulation. There are also very strongly controlled building height control planes, with precise definitions of exactly where you are allowed to build to. Evidently taking quite a bit of careful calculation at the planning stage of the process, the 45° setback planes to the boundary have had a considerable effect on the building planning. One side effect of this is that Dreicon have been equally careful and precise with the planning of the MEP risers in the building, avoiding any wasted space. Each riser will contain a separate zone for heating, for electrical, and for plumbing, all carefully laid out and meticulously implemented. Hot water is provided via a heat exchanger in the basement where the town-supplied hot water coils interact with the fresh water pipes of each building. The tight control of building services in this building, carefully planned beforehand, and the clustering of bathroom facilities and kitchens next to these services risers mean that space for services is kept to an absolute minimum.

District heating climate-friendly energy

According to the long-term strategy in Zurich's urban energy policy, climate-friendly energy sources are being further developed. The use of district heating contributes to the reduction of fossil fuels. Buildings that are connected to the district heating network do not need their own heating with fossil fuels. District heating in the city of Zurich is 80% CO_2-neutral because most of it comes from waste recycling and from wood. The optimal use of energy sources, combined with the most modern system technology and effective flue gas cleaning systems, guarantees maximum fuel utilisation and the best possible emission protection.

The district heating network

The district heating reaches the communal connections in the form of hot water via a well-insulated pipe network. The district heating network consists of two lines, the flow and the return, which form a closed water system. The flow brings the hot water from the heat production systems to the customers. Depending on the outside temperature, the water usually has temperatures of 90°C–120°C. The return line brings the water cooled to 50°C back to the heat production systems, which heat it up again, and the cycle starts all over again. The heat transfer at the communal connections takes place with the help of a transfer station: the heat is transferred to the common heating system of the connected buildings through heat exchangers.

Environmental benefits of district heating

Good for the climate:

The energy used is 80% CO_2-neutral and the majority of it is renewable.

Better air:

There are no more exhaust gases from heating systems in residential areas.

Local raw materials:

Thanks to local waste heat from waste and wood, the dependence on foreign energy is reduced.

Thermal insulation of buildings

The thermal insulation standard is officially regulated for the construction of new buildings in Switzerland. Part of the approval process is the creation of an energy certificate. The system verification can be calculated from the factors: heat transfer through the building envelope, heat generation, energy reference area and orientation of the building. The energy certificate must be submitted to the authorities for control and approval. The energy flow is calculated for all components of the building envelope. Both the winter and summer thermal insulation requirements must be met.

Handling of rainwater

The flat roof is designed as a green roof with high retention capacity. The sealing of the property is limited by the authorities. Rainwater infiltration into the ground allows for the maintenance of a constant groundwater level within each site as far as possible. The outer walls of the basement must be made correspondingly watertight. In Zurich, drainage lines for property drainage are only approved in exceptional cases.

Special features

Although common to houses in central Europe, a feature that may prove surprising to some is the use of retractable external shutters to the windows. This is a fairly common feature in Zurich – cleverly integrated into the external cladding so that a folded shutter can descend to provide privacy from summer heat and prying eyes. In a similar manner, the balconies here feature a track for a fabric curtain around each deck, again assuring privacy and heat relief.

More unusual perhaps, but a mandatory feature of all Swiss buildings is a secure, strengthened room in the basement. A hangover from the Cold War era many years ago, Switzerland has always remained neutral by being fearlessly well-prepared and well-armed, even today. Integrated into the basement storage areas behind the meticulously regimented car parking, a corner of the storage area can be converted quickly into a bunker in case of armed attack. Even more unusually perhaps, there also needs to be an escape tunnel excavated into the site, so that there is a way out from the bunker should the building be bombed to pieces. We're hoping that will never happen. The Swiss, along with the rest of the world, are working on what to build to handle the next Covid strain.

The new building looks very different from the previous pitched room version that was there before. Vertical, meticulously modern, and thoroughly efficient in its brand new construction system, yet comfortably nestling into the existing built-up neighbourhood, the new apartment building at 30 Regensbergstrasse is a vast improvement for all (Figures 21.3–21.11).

Figure 21.3 Regensbergstrasse typical apartment floor plan.
Source: Courtesy of Dreicon.

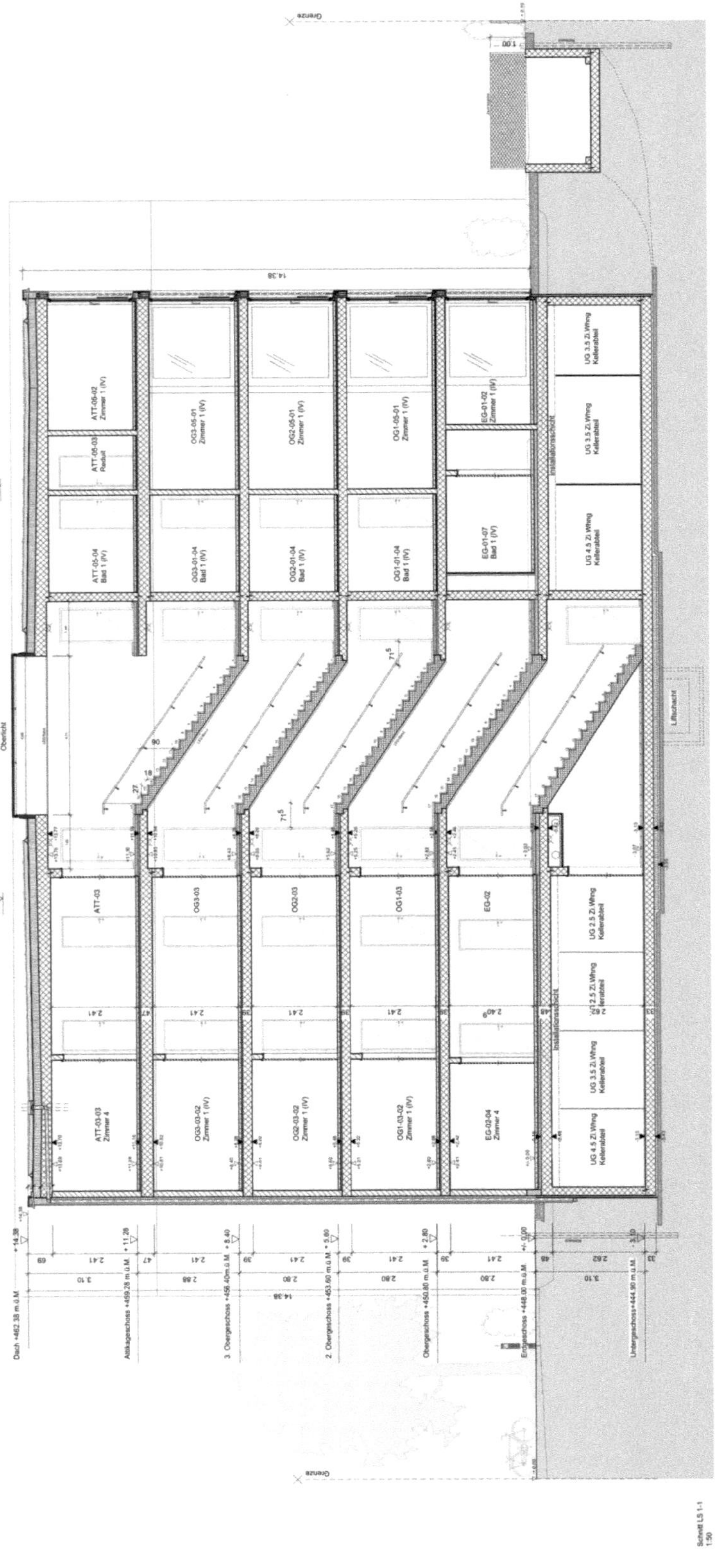

Figure 21.4 Regensbergstrasse cross-section.
Source: Courtesy of Dreicon.

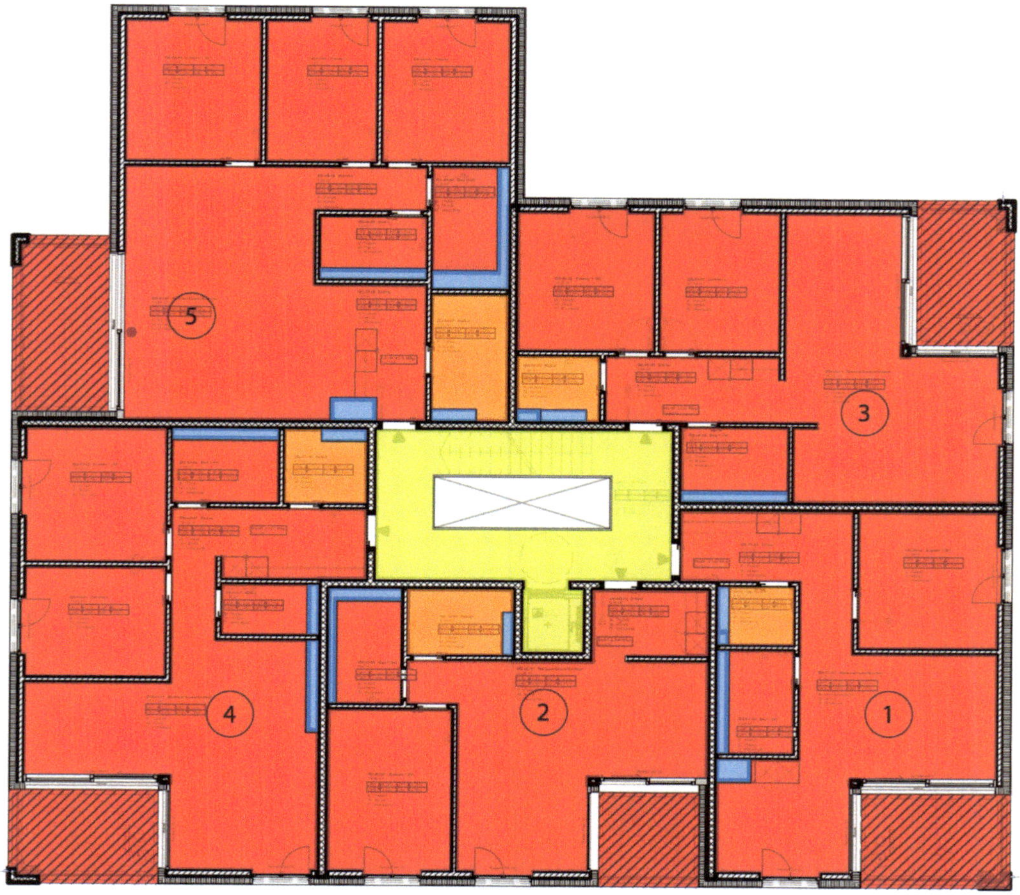

Figure 21.5 Regensbergstrasse planning diagram.
Source: Courtesy of Dreicon.

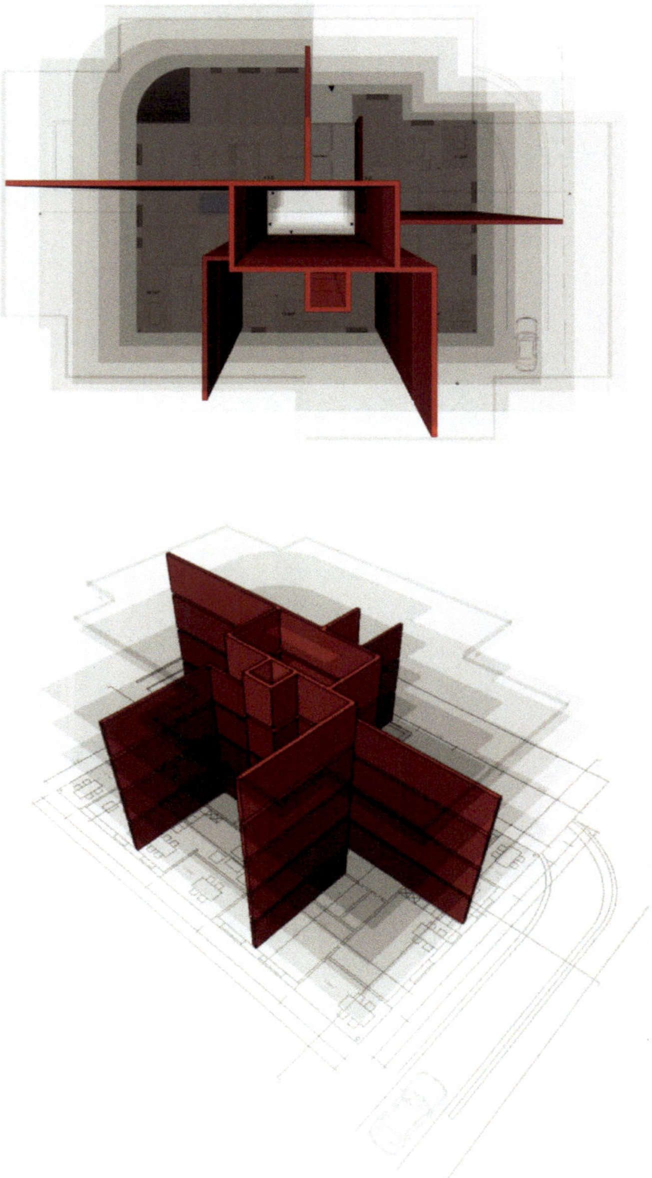

Figure 21.6 Regensbergstrasse structural diagrams.
Source: Courtesy of Dreicon.

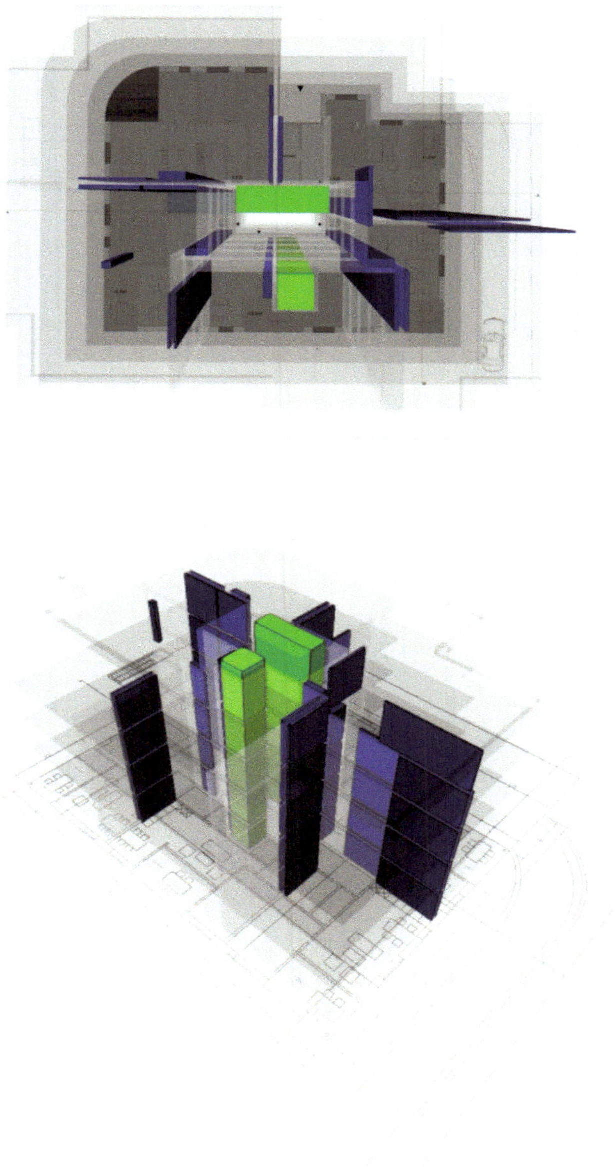

Figure 21.7 Regensbergstrasse services diagrams.
Source: Courtesy of Dreicon.

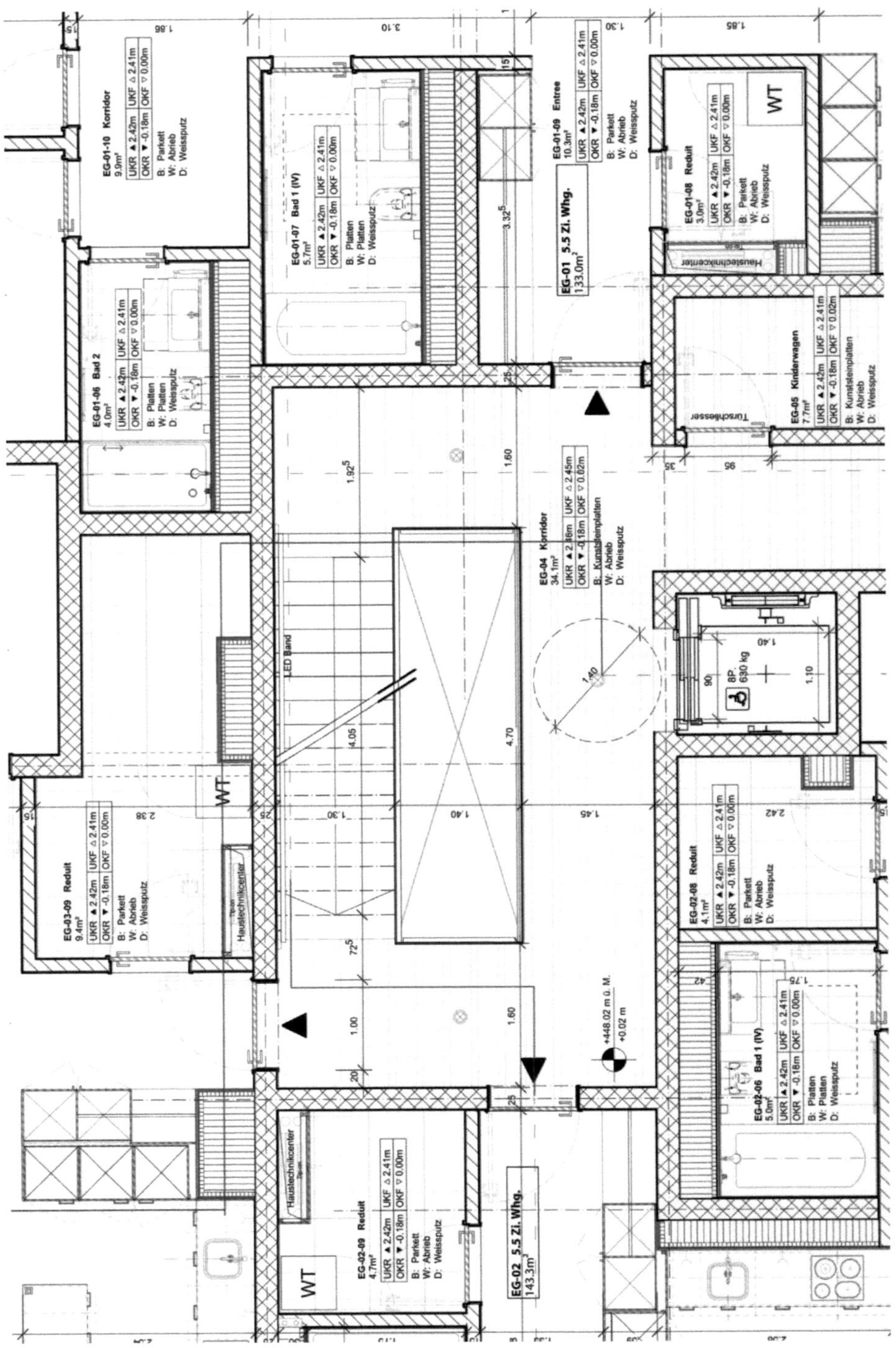

Figure 21.8 Regensbergstrasse apartment core planning.
Source: Courtesy of Dreicon.

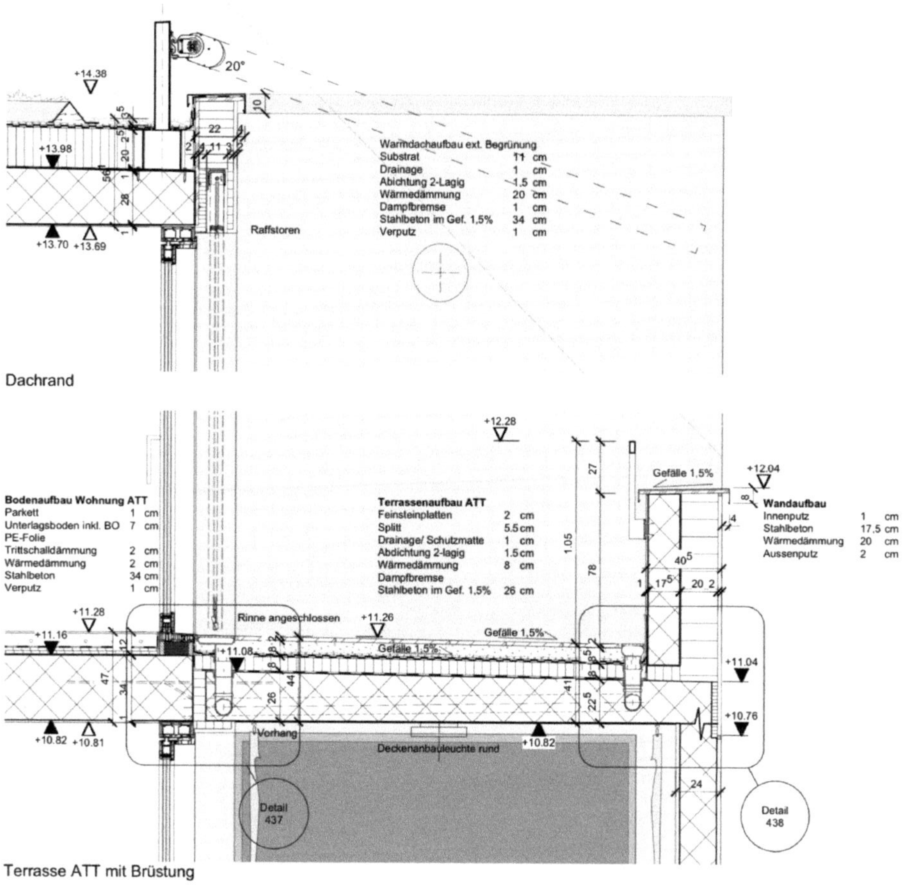

Figure 21.9 Regensbergstrasse section through apartment deck.
Source: Courtesy of Dreicon.

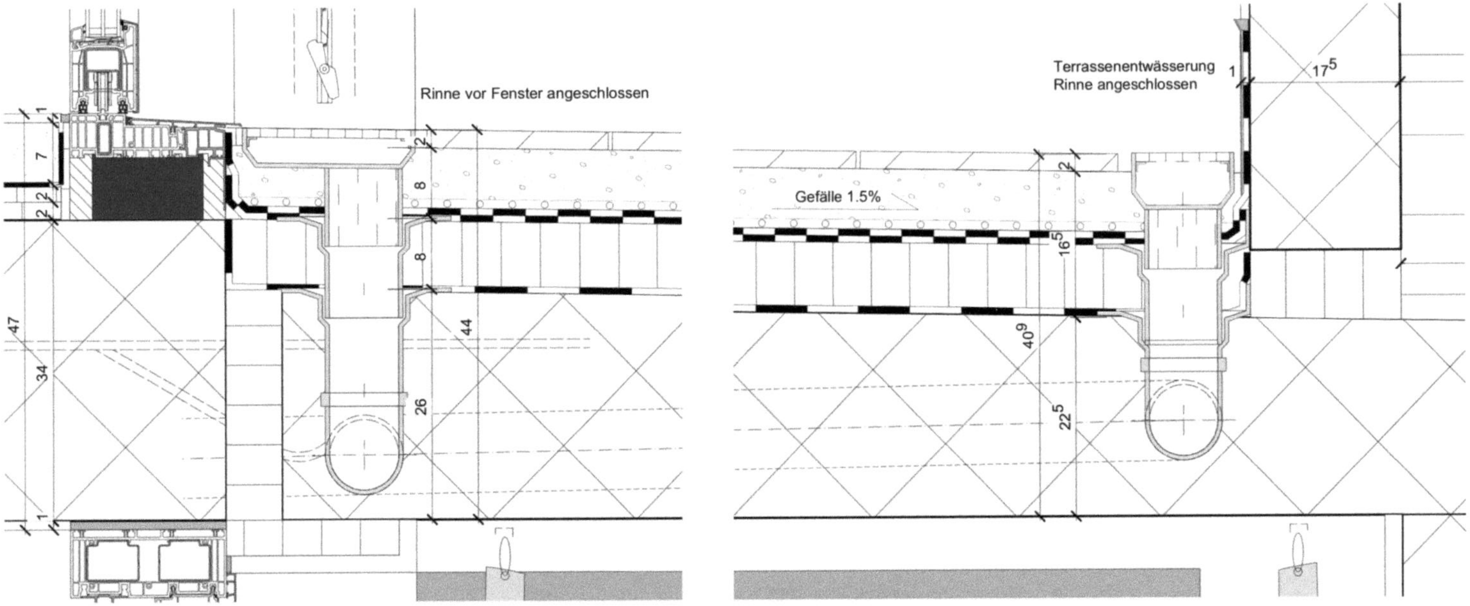

Detail 437 Anschluss Rinne ATT 1:5

Detail 438 Rinne ATT 1:5

Figure 21.10 Regensbergstrasse detail at apartment deck.
Source: Courtesy of Dreicon.

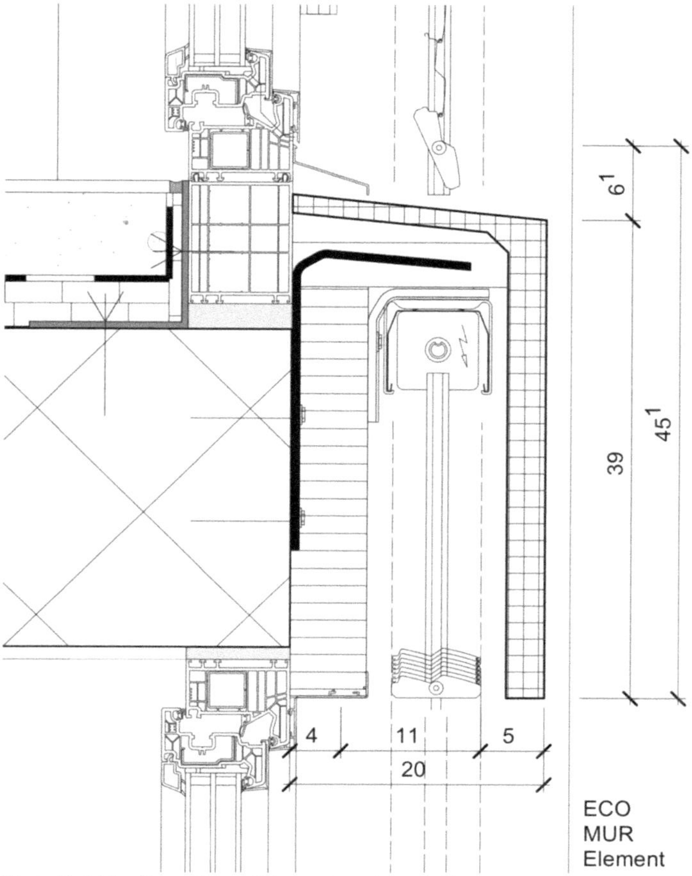

Detail 411 Fenster-/ Storenkasten 1:5

Figure 21.11 Regensbergstrasse external louvre detail.
Source: Courtesy of Dreicon.

Figures

Tables

Bibliography

A

Abrahamsen, Rune B. 2015. "First 14-storey wood building in the world at Bergen in Norway." *5th Formu International Bois Construction FBC.* Nancy: Forum Bois Construction, France.

Acton Ostry Architects and Fast + Epp: Brock Commons, Tallest Modern Timber Hi-Rise. Available at: https://clean50.com/projects/acton-ostry-architects-fast-epp-brock-commons-tallest-modern-timber-hi-rise/

Adams, M., Burrows, V., and Richardson, S. 2019. "Bringing embodied carbon upfront: Coordinated action for the building and construction sector to tackle embodied carbon." Report. London, United Kingdom: World Green Building Council.

Agrawal, A.W. and Schimek, P. 2007. "Extent and correlates of walking in the USA." *Transportation Research Part D: Transport and Environment, 12*(8), 548–563.

Alexander, C. 1977. *A Pattern Language: Towns, Buildings, Construction.* Oxford: Oxford University Press.

Alpern, A. 1975. *New York's Fabulous Luxury Apartments: With Original Floor Plans from the Dakota, River House, Olympic Tower and Other Great Buildings.* New York: McGraw-Hill.

Alread, J. and Leslie, T. 2007. *Design-Tech: Building Science for Architects.* Oxford, MA: Elsevier.

Ambrose, J. 2020. "Good vibrations: bladeless turbines could bring wind power to your home." *The Guardian*, 16 March 2021. Available at: https://www.theguardian.com/environment/2021/mar/16/good-vibrations-bladeless-turbines-could-bring-wind-power-to-your-home

Andregg, M. 2012. *Seven Billion and Counting: The Crisis in Global Population Growth.* Twenty-First Century Books.

Astel Modular. 2019. "Astel alustab Soome ühe suurima moodulitest korterelamuprojekti tootmisega (Astel to begin production for one of Finland's biggest modular apartment building projects)." *Astel Modular.* Accessed 27 March 2021. Available at: https://astelmodular.ee/astel-alustab-soome-uhe-suurima-moodulitest-korterelamuprojekti-tootmisega/

Atash, F. 1994. "Redesigning suburbia for walking and transit: emerging concepts." *Journal of Urban Planning and Development, 120*(1), 48–57.

Auckland Council. 2018. *Auckland Design Manual.* Auckland: Auckland Council.

B

Ballard, J.G. 1975. *High-Rise.* London: Penguin.

Barrett, D. 1998. *The Renewal of Trust in Residential Construction.* Canada: Commission of Inquiry into the Quality of Condominium Construction in British Columbia. Available at: http://www.qp.gov.bc.ca/condo/c3_iii.htm

BBC News, 2017. "Children interrupt BBC News interview – BBC News" 10 March 2017. Available at: https://www.youtube.com/watch?v=Mh4f9AYRCZY

Bell, P. 2009. "Kiwi Prefab." Master of Architecture thesis. Victoria University of Wellington, NZ.

Bell, P. and Southcombe, M. 2010. *Kiwi Prefab: cottage to cutting edge: prefabricated housing in New Zealand.* Auckland: Balasoglou Books.

Bentley, I. 1985. *Responsive Environments: A Manual for Designers.* London: Routledge.

Berghauser Pont, M.Y. and Haupt, P.A. 2009. "Space, density and urban form." Doctoral thesis.

Bhat, C.R. and Sen, S. 2006. Household vehicle type holdings and usage: an application of the multiple discrete-continuous extreme value (MDCEV) model. *Transportation Research Part B: Methodological, 40*(1), 35–53.

Biddulph, M. 2007. *Introduction to Residential Layout.* London: Routledge.

BIG, 2017. ARC CopenHill/Amager Bakke. Available at: https://big.dk/#projects-arc

Bolleter, J. and Ramalho, C.E. 2019. *Greenspace-Oriented Development: Reconciling Urban Density and Nature in Suburban Cities.* Cham: Springer.

Bonner, M., Wegrzynski, W., Papis, B., and Rein, G. February 2020. "KRESNIK: a top-down, statistical approach to understand the fire performance of building facades using standard test data." *Building and Environment, 169*, 106540. https://doi.org/10.1016/j.buildenv.2019.106540. Available at: https://www.sciencedirect.com/science/article/abs/pii/S0360132319307528?via%3Dihub

Bricker, D. and Ibbitson, J. 2019. *Empty Planet: The Shock of Global Population Decline.* London: Robinson.

Bristol, K. May 1991. "The Pruitt-Igoe myth." *Journal of Architectural Education, 44*(3), 163–171. Association of Collegiate Schools of Architecture. Available at: http://www.pruitt-igoe.com/temp/1991-bristol-pruitt-igoemyth.pdf

British Standards: Minimum gross internal floor areas sizes. 2015. *Technical Housing Standards – Nationally Described Space Standard*. Department for Communities and Local Government.

Broadbent, G. 2003. *Emerging Concepts in Urban Space Design*. London: Taylor & Francis.

Buchanan, A., Deam, B., Fragiacomo, M., Pampanin, S., and Palermo, A. 2008. "Multi-storey prestressed timber buildings in New Zealand." *Structural Engineering International, Journal of the International Association for Bridge and Structural Engineering (IABSE), 18*, 2.

Buchanan, A.H. and Fairweather, R.H. 1993. "Seismic design of glulam structures." *Bulletin of the New Zealand Society for Earthquake Engineering, 26*(4), 415–436. Available at: http://www.nzsee.org.nz/db/Bulletin/Archive/26(4)0415.pdf

Bundesministerium der Justiz und für Verbraucherschutz. n.d. *Gesetz über Ordnungswidrigkeiten (OWiG) 117Unzulässiger Lärm*. Germany: Bundesamt für Justiz. Available at: https://www.gesetze-im-internet.de/owig_1968/__117.html

Burgon, J. 1846. *Petra, a Poem: To Which a Few Short Poems Are Now Added* (2nd Edition. Oxford: MacPherson.

Busquets, J., Yang, D., and Keller, M. 2019. *Urban Grids: Handbook for Regular City Design*. Harvard: Oro Editions.

C

Cace, J., ter Horst, E., Syngellakis, K., Niel, M., Clement, P., Heppener, R., and Peirano, E. 2007. "Urban wind turbines – guidelines for small wind turbines in the urban environment." *Intelligent Energy Europe*. Available at: http://www.urban-wind.net/pdf/SMALL_WIND_TURBINES_GUIDE_final.pdf

Calthorpe, P., Fulton, W., and Fulton, W.B. 2001. *The Regional City*. Island Press, USA.

Carmona, M., Heath, T., Tiesdell, S., and Oc, T. 2010. *Public Places, Urban Spaces: The Dimensions of Urban Design*. London: Routledge.

Cattanach, A. and Clarke, R. 2020. "Scion hub: a three storey seismically resilient timber diagrid." *NZSEE 2020 Annual Conference*. Wellington: New Zealand Society for Earthquake Engineering.

Cattanach, A. and Pampanin, S. 2008. "21st century precast: the detailing and manufacture of New Zealands first multi-storey PRESSS building." *Proceedings of the 2008 New Zealand Concrete Industry Conference*, Rotorua, New Zealand, pp. 2–4.

CBS News. 2003. "Without warning, chaos. Who's to blame for a Chicago porch collapse that killed 13 people?" *CBS News*, 29 June 2003. Chicago:

13 people killed, 57 injured.Available at: https://www.cbsnews.com/news/without-warning-chaos/

CBS News. 2015. "Deadly balcony collapse tied to rotted wooden beams." *10TV WBNS*, 17 June 2015. California: 6 people dead. Available at: https://www.10tv.com/article/deadly-balcony-collapse-tied-rotted-wooden-beams

Cervero, R. and Kockelman, K. 1997. "Travel demand and the 3Ds: density, diversity, and design." *Transportation Research Part D: Transport and Environment, 2*(3), 199–219.

Cheng, V. 2010. "Understanding density and high density." In *Designing High-Density Cities*, 37–51. London: Routledge.

Cline, E. 2014. *1177 BC: The Year Civilization Collapsed.* Princeton, NJ: Princeton University Press.

CLT Handbook USA. *Think Wood.* Available at: https://info.thinkwood.com/clt-handbook

Cohen, S. and Sherrod, D.R. 1978. "When density matters: environmental control as a determinant of crowding effects in laboratory and residential settings." *Journal of Population, 1*(3), 189–202.

Coxworth, B. 2014. "Screwy-looking wind turbine makes little noise and a big claim." *New Atlas*, 27 May 2014. Available at: https://newatlas.com/the-archimedes-liam-f1-urban-wind-turbine/32263/

Cullen, G. 1995. *The Concise Townscape.* London: Routledge.

D

Davidson, H. 2019. "South Korea: world championship athletes injured in fatal balcony collapse." *The Guardian*, 27 July 2019. South Korea: 2 people dead, 16 injured. Available at: https://www.theguardian.com/world/2019/jul/27/south-korea-balcony-collapse-deaths-world-championship-swimmers-injured-

Defoe, D. 1722. *A Journal of the Plague Year – Being Observations or Memorials, of the Most Remarkable Occurrences, as well Publick as Private, Which Happened in London during the Last Great Visitation in 1665.* London: Penguin Classics (2003).

Department for Communities and Local Government. 2015. *Technical Housing Standards – Nationally Described Space Standard.* London: Department for Communities and Local Government

Department of Internal Affairs. 2019. Glossary. Available at: https://www.dia.govt.nz/diawebsite.nsf/wpg_URL/Resource-material-Building-Sustainable-Urban-Communities-Glossary

Ding, C., Liu, C., Zhang, Y., Yang, J., and Wang, Y. 2017. "Investigating the impacts of built environment on vehicle miles traveled and energy consumption: differences between commuting and non-commuting trips." *Cities, 68*, 25–36.

Donn, M. and Thomas, G. 2001. *Designing Comfortable Homes: Guidelines on theuse of Glass, Mass and Insulation for Energy Efficiency*. Wellington: EECA/CCANZ.

Dovey, K. and Pafka, E. 2014. "The urban density assemblage: modelling multiple measures." *Urban Design International, 19*(1), 66–76.

Dyer, P. 2019. *Rottenomics: The Story of New Zealand's Leaky Buildings Disaster*. New Zealand: Peter Dyer.

E

Ehasalu, A. 2015. "Rahvusvahelise turundusstrateegia arendamine Eesti Puitmajaklastri näited (Developing export marketing strategy: case of estonian wooden houses cluster)." Master's thesis. Tartu: Tartu University.

Enterprise Estonia. 2019. *Enterprise Estonia to support ten projects of professional associations*. 12 March. Accessed 28 March 2021. Available at: https://www.eas.ee/enterprise-estonia-to-support-ten-projects-of-professional-associations/?lang=en

Environment, M. f. 2011. *Medium-Density Housing in New Zealand*. Available at: https://www.mfe.govt.nz/more/towns-and-cities/medium-density-housing

ERR. 2019. *Keskkonnamaja võidutöö näitab puidu kasutust parimal moel (Environment House Winning Entry Shows Timber Used in the Best Possible Way)*. 20 March. Accessed 28 March 2021. Available at: https://kultuur.err.ee/921787/keskkonnamaja-voidutoo-naitab-puidu-kasutust-parimal-moel

Estonian Woodhouse Association. 2020. *Koroonaeelne aasta oli puitmajatootjate ekspordile edukas (The Year before Corona Was Successful for Timber House Producers)*. 24 August. Accessed 27 March 2021. Available at: https://puitmajaliit.ee/uudised/koroonaeelne-aasta-oli-puitmajatootjate-ekspordile-edukas

Ewing, R. and Hamidi, S. 2017. *Costs of Sprawl*. London: Taylor & Francis.

F

Frey, H. 2003. *Designing the City: Towards a More Sustainable Urban Form*. London: Taylor & Francis.

Frye, D. 2018. *Walls: A History of Civilisation in Blood and Brick*. London: Faber & Faber.

G

Gelfand, M. 2020. "Why countries with 'loose', rule-breaking cultures have been hit harder by Covid." *The Guardian*, 1 February 2020. Available at: https://www.theguardian.com/world/commentisfree/2021/feb/01/loose-rule-breaking-culture-covid-deaths-societies-pandemic

Gifford, R. 2007. "The consequences of living in high-rise buildings." *Architectural Science Review, 50*(1), 2–17.

Glaeser, E. 2011. *Triumph of the City: How Urban Spaces Make Us Human.* London: MacMillan.

Glancey, J. 2010. "Spin city: Strata Tower." *The Guardian*, 18 July 2010. Available at: https://www.theguardian.com/artanddesign/2010/jul/18/strata-tower-london-green-architecture

Golany, G.S. 1996. "Urban design morphology and thermal performance." *Atmospheric Environment, 30*(3), 455–465.

Green, M. 2018. *The Case for Tall Wood Buildings* (2nd Edition). Seattle, WA: Michael Green Architecture.

Green, M. and Taggart, J. 2017. *Tall Wood Buildings – Design, Construction and Performance.* Basel: Birkhauser.

Grenfell Tower Inquiry. 2019. *Grenfell Tower Inquiry: Phase 1 Report.* The Rt Hon Sir Martin Moore-Bick. London: Her Majesty's Stationery Office.

H

Heathcote, C. May 2020. "When modernist Britain fell down." In *Quadrant*, 82. Sydney: Quadrant.org.au.

Heckmann, O. and Schneider, F. (eds). 2011. *Floor Plan Manual Housing* (4th Edition). Basel: Birkhauser.

Hempel, S. 2006. *The Medical Detective: John Snow, Cholera, and the Mystery of the Broad Street Pump.* London: Granta Books.

Hess, D. B. and Tiit T. 2019. "Modernist housing estates in the baltic countries: formation, current challenges and future prospects." In *Housing Estates in the Baltic Countries: The Legacy of Central Planning in Estonia, Latvia and Lithuania*, edited by Daniel Baldwin Hess and Tiit Tammaru, 3–30. Cham: Springer.

Hong Kong Government of Planning Department. 1994. *Hong Kong Planning Standards and Guidelines.*

I

IPCC. 2018. 'Global warming of 1.5°C.' Special Report.

Ismael, H.M. 2020. "Urban form study: the sprawling city—review of methods of studying urban sprawl." *GeoJournal, 86*(2) 1–12

J

Jabareen, Y.R. 2006. "Sustainable urban forms: their typologies, models, and concepts." *Journal of Planning Education and Research, 26*(1), 38–52.

Jacobs, A. and Appleyard, D. 1987. "Toward an urban design manifesto." *Journal of the American Planning Association, 53*(1), 112–120.

Jacobs, J. 2016. *The Death and Life of Great American Cities.* New York: Vintage.

Jarosz, B. 2017. "Condo balcony collapse prompts 150+ decks deemed 'unsafe'." *ABC 6*, 10 June 2017. Hayden Run, Colorado: 2 people injured. Available at: https://abc6onyourside.com/news/local/condo-balcony-collapse-prompts-150-decks-deemed-unsafe-
06-10-2017

Jones, S. 2019. *Mass Timber – Design and Research.* Novato, CA: ORO Editions.

K

Kallas, J. 2018. "Innovatsioon ja selle soodustamise viisid Eesti puitmajatootjate näitel (Innovation and the ways to stimulate it at the example of estonian wood house producers)." Master's thesis. Tartu: Tartu University.

Kaufmann, H., Krotsch, S., and Winter, S. 2018. *Manual of Multi-Storey Timber Construction.* Munich: Detail.

Kearns, A., Whitley, E., Mason, P., and Bond, L. 2012. "'Living the high life'? Residential, social and psychosocial outcomes for high-rise occupants in a deprived context." *Housing Studies, 27*(1), 97–126.

Kenway, P. and Holden, J. 11 April 2020. *Accounting for the Variation in the Confirmed Covid-19 Caseload across England: An Analysis of the Role of Multi-Generation Households, London and Time.* New Policy Institute (NPI). Available at: https://www.npi.org.uk/files/2115/8661/6941/20-04-11_Accounting_for_the_variation_in_Covid_cases_across_England.pdf

Kleerekoper, L., van Esch, M., and Salcedo, T.B. 2012. "How to make a city climate-proof, addressing the urban heat island effect." *Resources, Conservation and Recycling, 64*, 30–38.

KLH. n.d. a. *Panel Types, Surfaces & Technical Characteristics.* Available at: https://www.klh.at/en/panel-types-surfaces-technical-details/#technical-details

KLH. n.d. -b. *Component Catalogue for Cross Laminated Timber Structures.* Austria: KLH Massivholz GMBH. Available at www.klh.at/en/klh-construction-en.pdf

Kodumaja, A.S. 2019. "Kodumaja Presentatsioon / Kodumaja presentation." *Kodumaja.* 27 May. Accessed 27 March 2021. http://www.kodumaja.ee/et/Presentatsioonid.

Kremer, P. *LinkedIn.* Available at: https://www.linkedin.com/pulse/designing-mass-timber-what-dont-teach-you-paul-kremer/

Kuusk, K., Phelo, P., and Kalamees, T. 2019. "Renovation of apartment buildings with prefabricated modular panels." *REHVA 13th HVAC World Congress CLIMA 2019.* Bucharest: E3S Web of Conferences 111.

L

Laar, M. 2007. "The Estonian economic miracle." *Backgrounder* (The Heritage Foundation), 2060, 1–12.

Lane, B. M. 2006. *Housing and Dwelling: Perspectives on Modern Domestic Architecture.* London: Routledge.

Lasner, M. G. 2012. *High Life: Condo Living in the Suburban Century.* New Haven, CT and London: Yale University Press.

Le Camus de Mezieres, N. 1992. *The Genius of Architecture; or, the Analogy of that Art with Our Sensations.* Translated by David Britt. Santa Monica: The Getty Centre.

Le Corbusier, J. 1928. *Corbusier Le Grand.* Paris.

Leslie, T. 2013. *Chicago Skyscrapers (1871–1934).* Urbana, Chicago, and Springfield: University of Illinois Press.

LETI. 2020. *LETI Climate Emergency Design Guide.* London: London Energy Transformation Initiative.

Levitt, D. and McCafferty, J. 2019. *The House Design Handbook: A Guide to Good Practice* (2nd Edition). New York: Routledge.

Liotta, P.H. and Miskel, J. 2012. *The Real Population Bomb: Megacities, Global Security & the Map of the Future.* Washington DC: Potomac Books.

Lipstadt, H. 1977. "Housing the Bourgeoisie: Cesar Daly and the ideal home." *Oppositions, 8* (Spring), 41.

Litman, T. and Steele, R. 2017. *Land Use Impacts on Transport.* Victoria Transport Policy Institute, Canada.

Livi-Bacci, M. 2017. *A Concise History of World Population* (6th Edition). London: Wiley Blackwell.

Lufkin, B. 2020. "What outdoor space tells us about inequality." 18 June 2020. *BBC Worklife.* Available at: https://www.bbc.com/worklife/article/20200617-the-luxury-and-privilege-of-a-balcony-or-yard-during-covid

Lumiste, R. Pefferly, R. and Purju, A. 2008. *Estonia's Economic Development: Trends, Practices and Sources.* Case Study. Washington, DC: The International Bank for Reconstruction and Development/The World Bank.

Lynch, K. 1984. *Good City Form.* Cambridge Massachusetts: MIT Press.

M

Maas, J., Verheij, R.A., Groenewegen, P.P., De Vries, S., and Spreeuwenberg, P. 2006. "Green space, urbanity, and health: how strong is the relation?" *Journal of Epidemiology & Community Health, 60*(7), 587–592.

Madsen, J. 2019 "A Danish fiasco: the Copenhagen incineration plant." *Zero Waste Europe.* Available at: https://zerowasteeurope.eu/2019/11/copenhagen-incineration-plant/

Manaugh, G. 2016. *A Burglar's Guide to the City.* New York: Farrar, Strauss and Giroux.

Marcus, S. 1999. *Apartment Stories: City and Home in Nineteenth Century Paris and London*, 19. Berkeley: University of California Press.

Marshall, S. 2004. *Streets and Patterns.* London: Routledge.

Matek. 2021. *Slottet.* Accessed 27 March 2021. Available at: https://www.matek.ee/et/project/slottet/

Mattson, J. 2020. "Relationships between density, transit, and household expenditures in small urban areas." *Transportation Research Interdisciplinary Perspectives, 8*, 100260.

Marriage, G. May/June 2003. "No swinging cats in Inner-City flats." *Architecture New Zealand.* Auckland: AGM Publishing.

Marriage, G. 2010. "Minimum vs maximum: size and the New Zealand house." In *2010 Australasian Housing Researchers' Conference.* Auckland: University of Auckland.

Marriage, G. 2014. "Pavlova paradise and the curse of the side yard." In *Building a Better New Zealand Conference.* Auckland: MBIE.

Marriage, G. 2019. *Tall: The Design and Construction of High-Rise Architecture.* London: Routledge.

Marriage, G. and Sutherland, B. 2014. *New Digital Housing Typologies: CNC Fabrications of CLT Structure and BIM Cladding.* Architectural Science Association: ANZASCA.

Mathieson, K. 2015. "Where is the world's windiest city? Spoiler alert: it's not Chicago." *The Guardian*, 15 October 2015. Available at: https://www.theguardian.com/cities/2015/oct/15/where-world-windiest-city-spoiler-alert-chicago-wellington

Mayo, J. 2015. *Solid Wood – Case Studies in Mass Timber Architecture, Technology and Design.* London: Routledge.

McNeilly, H. 2017. "Multiple failings revealed in Dunedin Six60 gig balcony collapse." *Stuff*, 3 February 2017. Dunedin, New Zealand: 18 injured, 5 seriously, with one person having severe spinal injuries. Available at: https://www.stuff.co.nz/national/89073303/multiple-failings-revealed-in-dunedin-six60-gig-balcony-collapse

Merrick, J. 2017. "High density, low carbon: Dalston Works by Waugh Thistleton." *Architects Journal*, 4 October 2017. Available at: https://www.architectsjournal.co.uk/buildings/high-density-low-carbon-dalston-works-by-waugh-thistleton

Metslang, J. 2016. "Vanemad palktarandid/Development of wooden constructions." rmt: *Puit Eesti arhitektuuris/Wood in Estonian Architecture*, Leele Välja, 17–42. Tallinn: Metsa- ja Puidutööstuse Liit.

Middleton, R. 1983. "I. The use and abuse of tradition in architecture." *Journal of the Royal Society of Arts, 131*(5328), 729–739.

Middleton, R. 1992. 'Introduction'. In *The Genius of Architecture; or, the Analogy of that Art with Our Sensations*, edited by Nicolas Le Camus de Mezieres, 32. Santa Monica, CA: The Getty Centre for the History of Art and the Humanities.

Ministry of Housing, Communities and Local Government (MHCLG) Joint Industry Working Group on MMC. 2019. *Developing a Definition Framework.* Working Group members include Build offsite, Homes England, National House Building Council (NHBC) and Royal Institution of Chartered Surveyors (RICS). Available at: https://www.gov.uk/government/publications/modern-methods-of-construction-working-group-developing-a-definition-framework

Mitt, A. 2016. "Eesti Puitmajatootjate Tarneahelad Välisturule Sisenimisel (Estonian wooden house manufacturers supply chains for entering foreign market)." Master's thesis. Tallinn: Tallinn Technical University.

Morris, P. 1961. *Homes for Today and Tomorrow.* London: Her Majesty's Stationery Office.

Moughtin, C. 2003. *Urban Design: Street and Square*. London: Routledge.

Moussavi, F. 2014. *The Function of Style*. New York: Actar/Harvard University Graduate School of Design/Function Lab.

Murray, J. 2020. "Manchester tower block residents ineligible for £1bn recladding fund." *The Guardian*, 31 May 2020. Available at: https://www.theguardian.com/uk-news/2020/may/31/manchester-tower-block-residents-ineligible-1bn-recladding-fund-skyline-central

N

Naylor Love. 2021. *Building Carbon Calculator*. Available at: https://www.naylorlove.co.nz/carbon/

Newcombe, M.P., Pampanin, S., Buchanan, A., and Palermo, A. 2008. "Section Analysis and Cyclic Behavior of Post-Tensioned Jointed Ductile Connections for Multi-Storey Timber Buildings." *Journal of Earthquake Engineering, 12*(Special Issue), 83–110.

Newcombe, M. 2011. "Seismic design of post-tensioned timber frame and wall buildings." PhD thesis. University of Canterbury, NZ.

Newman, O. 1961. *New Frontiers in Architecture*. A summary of the 1959 International Congress of Modern Architects conference in Otterlo. An early exploration of the effects of architectural design on perception and behavior. New York: Universe Books.

Newman, O. 1996. *Creating Defensible Space*. New York: U.S. Department of Housing and Urban Development Office of Policy Development and Research. Available at https://www.huduser.gov/Publications/pdf/def.pdf

Newman, O. 1972. *Defensible Space*. New York: Macmillan.

Newman, P.W. and Kenworthy, J.R. 1989. "Gasoline consumption and cities: a comparison of US cities with a global survey." *Journal of the American Planning Association, 55*(1), 24–37.

Ng, E., Chen, L., Wang, Y., and Yuan, C. 2012. "A study on the cooling effects of greening in a high-density city: an experience from Hong Kong." *Building and Environment, 47*, 256–271.

Niu, J. 2004. "Some significant environmental issues in high-rise residential building design in urban areas." *Energy and Buildings, 36*(12), 1259–1263.

O

Omrani, S., Garcia-Hansen, V., Drogemuller, R., and Capra, B. 2017. "On the effect of provision of balconies on natural ventilation and thermal comfort in high-rise residential buildings." *Building and Environment, 123*, 504–516. https://doi.org/10.1016/j.buildenv.2017.07.016

Omrani, S., Garcia-Hansen, V., Drogemuller, R., and Capra, B. 2016. "Thermal comfort evaluation of natural ventilation mode: case study of a high-rise residential building." In *Fifty Years Later: Revisiting the Role of Architectural Science in Design and Practice: 50th International Conference of the Architectural Science Association 2016*, edited by J. Zuo, L. Daniel, and V. Soebarto, 1–9. Adelaide, SA: The Architectural Science Association and the University of Adelaide.

P

Pagett, N. and Smit, J. 2013. *The End of More: Infinite Demand on Finite Resources Is Making Humankind Unsustainable*. London: Lees Publishing

Palermo, A., Pampanin, S., Buchanan, A., and Newcombe, M. 2005. "Seismic design of multi-storey buildings using laminated veneer lumber (LVL)." In *2005 New Zealand Society of Earthquake Engineering Conference*, Wairaki, New Zealand, p. 8.

Parker, J. 2017. "Mass appeal." Architects data file.
Available at: https://architectsdatafile.co.uk/news/mass-appeal-brock-commons-tallwood-house-vancouver-canada/.

Pearson, A. 2016. "Dalston lane: tall timber." *Building*, 20 January 2016. Available at: https://www.building.co.uk/dalston-lane-tall-timber/5079749.article

Peiser, R.B. 1989. "Density and urban sprawl." *Land Economics, 65*(3), 193–204.

Petrovic, E. 2003. "Your new house – or would a flat suit you better? Development of housing preferences in England, Paris, Germany and New Zealand." Master's thesis. Victoria University of Wellington.

Poon, L. 2020. "A lesson from social distancing: build better balconies." *Bloomberg CityLab*. 21 April 2020. Available at: https://www.bloomberg.com/news/articles/2020-04-20/lesson-from-coronavirus-build-better-balconies

Available https://Prefabnz.com

Q–R

Ramsdonk, T. 2020. *A New Zealand based Study on Airtightness and Moisture Management*. Wellington: Proclima. Available at: https://proclima.co.nz/pro-clima-study/

Rasmussen, S.E. 1964. *Experiencing Architecture* (Vol. 2). Cambridge, MA: MIT Press.

Riistop, M. 2016. "Uued puitkonstruktsioonid/New wooden constructions." rmt: *Puit Eesti arhitektuuris/Wood in Estonian Architecture*, Leele Välja, 43–54. Tallinn: Metsa- ja Puidutööstuse Liit.

Robust D. 2020. *Part E Robust Details October 2020 Revision.* Available at: https://www.robustdetails.com

Rodenburg, C., Vreeker, R., and Nijkamp, P (Eds). 2003. "Multifunctional land use: an economic perspective." In *The Economics of Multifunctional Land Use*, 3–15. Amsterdam: Shaker.

S

Schuler, T. 2018. "Dalston Works, the largest CLT building in the world." *Architect*, 9 October 2018. Available at: https://www.architectmagazine.com/technology/detail/dalston-works-the-largest-clt-building-in-the-world_o

Seto, K.C., Güneralp, B., and Hutyra, L.R. 2012. Global forecasts of urban expansion to 2030 and direct impacts on biodiversity and carbon pools. *Proceedings of the National Academy of Sciences, 109*(40), 16083–16088.

Shoenauer, N. 2000. *6,000 Years of Housing.* London: WW North and Co.

Sills van Bohemon Architects. 2005. *Good Solutions for Apartment Buildings.* Auckland: North Shore Council.

Silva, C.V., Branco, J.M., and Lourenço, P.B. 2013. "A project contribution to the development of sustainable multi-storey timber buildings." In *Portugal SB13*, Guimarães, Portugal.

Silva, C.V., Branco, J.M., and Lourenço, P.B. 2014. "UT system: a structural system to build taller urban timber houses with aspired spatial flexibility." In *40th IAHS World Congress on Housing*, Madeira Island, Portugal.

Smith, M. 2019. *Cities: The First 6,000 Years.* London: Simon & Schuster.

Smith, R. 2010. *Prefab Architecture: A Guide to Modular Design and Construction.* Hoboken: Wiley.

Stora Enso Wood Products. 2021. *Stora Enso CLT Technical Brochure.* Available at: https://www.storaenso.com/en/products/wood-products/massive-wood-construction/clt

Studio Pacific. n.d. *Appartamento il Casino.* Available at: https://www.studiopacific.co.nz/projects/2013/appartamento-il-casino/

Sturges, F. 2019. "Tate Modern's neighbours miss the point: cities make voyeurs of us all." *The Guardian*, 16 February 2019. Available at: https://www.theguardian.com/commentisfree/2019/feb/16/tate-modern-angry-neighbours-cities-voyeurs

Sturgis, S. 2017. *Targeting Zero: Embodied and Whole Life Carbon Explained.* Newcastle upon Tyne: RIBA Publishing.

Susilo, Y.O. and Stead, D. 2009. "Individual carbon dioxide emissions and potential for reduction in the Netherlands and the United Kingdom." *Transportation Research Record, 2139*(1), 142–152.

Sutcliffe, A. 1981. *Towards the Planned City: Germany, Britain, the United States and France, 1780–1914*, 116. Oxford: Basil Blackwell.

Sutherland, J. 2019. "Façade: design." In *Tall: the Design and Construction of High-Rise Architecture*, edited by Guy Marriage. London: Routledge, 275–314.

Swanwick, C., Dunnett, N., and Woolley, H. 2003. "Nature, role and value of green space in towns and cities: an overview." *Built Environment, (1978-)*, 94–106.

Swedish Wood magazine. 2020. "A global solution for a locally active industry." *Wood Magazine*, issue 2, 2020. Available at: https://www.swedishwood.com/publications/wood-magazine/2020-2/gerhard-schickhofer/

T

Tammur, A., Tammaru, T., and Puur, A. 2017. "Is there a migration turnaround taking place in Estonia? Migration trends 2000–2015." In *Estonian Human Development Report 2016/2017 Estonia at the Age of Migration*, edited by Raul Eamets, Kristina Kallas, and Tiit Tammaru. Tallinn: Foundation Estonian Cooperation Assembly, 3–17.

Taylor, D. 2020. "Homes in London high-rises left 'wet and cold' after cladding removed." *The Guardian*, 22 January 2020. Available at: https://www.theguardian.com/society/2020/jan/22/homes-london-high-rises-left-wet-cold-cladding-removed

Torres, A., Liu, J., Brandt, J., and Lear, K. 2017. "The world is running out of sand." *Smithsonian Magazine*, 8 September 2017. Available at:https://www.smithsonianmag.com/science-nature/world-facing-global-sand-crisis-180964815/

U

UN-Habitat. 2004. "The challenge of slums: global report on human settlements 2003." *Management of Environmental Quality: An International Journal, 15*(3), 337–338.

United Nations. 2015. "Paris Agreement."

V

Vaitmaa, E. 2016. "Tallinna Maja." *Forte*, 30 September. Available at: https://longread.delfi.ee/forte/tallinna-maja?id=75652667

Valk-Siska, V (ed.). 2019. *The Cutting Edge: Estonia's Leading Timber Construction Enterprises*, edited by Veronika Valk-Siska. Tallinn: Estonian Woodhouse Association.

Venturi, R., Stierli, M., and Brownlee, D.B. 1977. *Complexity and Contradiction in Architecture (Vol. 1): The Museum of Modern Art.*

W

Wainwright, O. 13 April 2020. "Smart lifts, lonely workers, no towers or tourists: architecture after coronavirus." *The Guardian.* Available at:https://www.theguardian.com/artanddesign/2020/apr/13/smart-lifts-lonely-workers-no-towers-architecture-after-covid-19-coronavirus

Walker, K. 2018. "Towards a construction revolution." Public address to *Timber Design Conference* in Lithuania.

Wall, T. 2020. "Cramped living conditions may be accelerating UK spread of coronavirus." *The Observer,* 12 April 2020. Available at: https://www.theguardian.com/world/2020/apr/12/virus-hitting-hardest-modern-equivalent-victorian-slums

Warburton, N. 2004. *Erno Goldfinger – The Life of an Architect.* Oxon: Routledge.

Waugh Thistleton Architects. n.d. *Dalston Works.* Available at: https://waughthistleton.com/dalston-works/

Waugh Thistleton Architects. 2018. *100 Projects UK CLT.* London: Softwood Lumber Board and Forestry Innovation Investment + Waugh Thistleton Architects.

Wilson, B. 2020. *Metropolis: A History of Humankind's Greatest Invention.* London: Jonathan Cape.

World Health Organisation, n.d. *Head Circumference-for-Age (Birth to 5 years).* WHO Child Growth Standards.

Wolch, J.R., Byrne, J., and Newell, J.P. 2014. "Urban green space, public health, and environmental justice: the challenge of making cities 'just green enough'." *Landscape and Urban Planning, 125,* 234–244.

X–Y

Yang, J. and French, S. 2013. "The travel—obesity connection: discerning the impacts of commuting trips with the perspective of individual energy expenditure and time use." *Environment and Planning B: Planning and Design, 40*(4), 617–629.

Yang, W., Li, T., and Cao, X. 2015. "Examining the impacts of socio-economic factors, urban form and transportation development on CO_2 emissions from transportation in China: a panel data analysis of China's provinces." *Habitat International, 49,* 212–220.

Z

Zuskin, R. 1999. *The Place Vendome: Architecture and Social Mobility in Eighteenth-Century Paris,* 65. Cambridge: Cambridge University Press.

Index

Taylor & Francis Group
an **informa** business